ATLAS
OF THE SLITLAMP-MICROSCOPY
OF THE LIVING EYE

TECHNIC AND METHODS OF EXAMINATION

BY

PROF. DR. ALFRED VOGT

DIRECTOR OF THE OPHTHALMIC HOSPITAL AT THE
UNIVERSITY OF BASLE

*

AUTHORIZED TRANSLATION BY
DR. ROBERT VON DER HEYDT, CHICAGO

*

WITH 370 ENGRAVINGS

Springer-Verlag Berlin Heidelberg GmbH
1921

DEDICATED TO

ALLVAR GULLSTRAND

PREFATORY NOTE

The illustrations shown in this atlas present observations made on patients of my clinic, with the exception of a few showing the suture areas and photographic representations of the tunica vasculosa lentis, which latter I have made from preparations obtained from various local University Institutes.

I am very much indebted to the artist, *Mr. Jak. Iseli*, who has made the greater number of illustrations with utmost skill, care and devotion to the subject.

I also wish to thank the translators for their extraordinary pains and devotion, especially Dr. *F. Ed. Koby*, first Assistant at the clinic, Prof. *Verderame* in Turin, and Dr. *R. von der Heydt* in Chicago. It was not an easy matter to find and apply fitting expressions for heretofore unknown conceptions. The choice of the technical terms necessary often taxed to the utmost the descriptive power of words.

For his kind assistance in the revision of the English text I am indebted to Prof. Dr. *H. K. Corning*, Director of the Institute of Normal Anatomy of our University, and to Dr. *A. Darier*, Editor of the Clinique Ophthalmologique, in Paris, for that of the French text.

In connection with the English translation of the Introduction and the chapter pertaining to the cornea Dr. med. *Harriet Parrell* of Basel has willingly participated and I owe her many thanks. Especial credit must be given the publishing firm of *Julius Springer* who have done such excellent work toward the success of this atlas. For this service I wish to here express my thanks and appreciation.

I fully recognize that this atlas in its present form is quite incomplete and that only a small part of the new observations made possible by the combination of the slitlamp with the corneal microscope have been as yet made accessible. It however will give a fundamental idea as to the great importance of this new subject and will no doubt inspire to further investigations.

It should be expected that the method of application in so new a field of research is at first of greater importance than routine of practice. Instructions in the method of application and technic have not as yet been disclosed and in this respect I hope this work will be of aid.

A few of the methods of investigation, for instance, the use of reflecting zones, as well as some of the drawings in the atlas, have been recently published in various ophthalmological journals; the majority, however, are new and all are now systematically classified. In executing the illustrations special care was taken to give reproductions true to nature. Sketches and schematic presentations were avoided as much as possible.

BASEL, June 1921   A. VOGT

CONTENTS

Remarks referring to the illustrations.

The original drawings are, on the whole, well represented. A few specially fine details are however somewhat indistinct, for instance, the illustrations of corneal endothelium, corneal dew-like changes, folds in the lens capsule, pigment dots.

In some of the illustrations an incorrect, pinkish-red tint appears, for instance, in Fig. 140a, 197, 199, 211, 217a, 257, 316, 359b. The pigment dots are often too red (instead of brown), for instance, in Fig. 270. In Fig. 47 and 48 the pigment line should be olive-coloured instead of reddish-brown, about the same as in Fig. 46a Fig. 140a is spotty,

In Fig. 257 the tint should be the same as that in Fig. 256.

A.
INTRODUCTION

The introduction of the slitlamp of Gullstrand has opened to ophthalmology an entirely new field for clinical observation and diagnosis.

It has created what may be termed an "Histology of the living eye".

Normal and pathologic conditions heretofore established only anatomically can be directly observed in the living eye. It shows us not only structures that are known, but in addition a series of observations on histological details, heretofore impossible. These structures, owing partly to their delicacy, were formerly sacrificed in the process of fixation, or it was impossible to further differentiate them by any method of staining. For instance, we have up to the present failed of anatomical proof of the numerous physiologic remnants of the tunica vasculosa lentis, the arteria hyaloidea, the various intricacies of the framework supporting the vitreous body, types of lens sclerosis, etc., but the number of facts already known as a result of anatomical research which have hitherto evaded clinical confirmation, is far greater.

The slitlamp, in combination with the corneal microscope, permits us to observe the living endothelium on the posterior surface of the cornea. Every individual endothelial cell on Descemet's membrane, as well as each pathologically deposited lymphocyte is revealed. The nerve fibres of the cornea can be traced to their very finest ramifications. In Descemet's and Bowman's membranes we have observed pathological folds, manifested by their characteristic reflexes. We can see the blood corpuscles as they roll along in newly formed corneal vessels, as well as within the vessels that form the vascular loops at the limbus.

Oedema of the corneal endothelium or epithelium is indicated by a deposit like fine dew. The slitlamp has disclosed or explained a multitude of pathological conditions hitherto unknown or obscure.

More important than these details in the cornea have been the investigations which the method enabled us to make of the development of the lens. It disclosed the paucity of our knowledge regarding the genesis and morphology of senile cataract.

A multitude of clinically unknown manifestations in cataract formation are revealed. We learn to recognize the form of subcapsular striae of vacuole formation in advanced cataract, the folding or wrinkling of the capsule when shrinking begins, the so often unobserved variety of types of nuclear sclerosis, the peripheral concentric lamellar clouding, the various types of wreath and rosette shaped cataract, the genesis of spokes and cystic spaces, the characteristic picture of lamellar separation, of posterior cortical cataract, etc.

The slitlamp teaches us to differentiate the acquired from the various forms of congenital cataract, it also gives us the first definite clinically sharp point of differentiation between cataracta complicata and cataracta senilis. Our knowledge of the physiology of the normal youthful and the ageing lens is enriched beyond expectation by the new method. We can see the lens epithelium and the anterior and posterior graining (shagreen) in its normal and changed condition. The maximum of the interior reflection of the lens, its lamination, of which latter we had only a very vague conception, based on our knowledge of the formation of the nucleus, and of which we were in part in total ignorance, now appear revealed before our eyes and can be traced with the bundle of light as to their form, arrangement, number and degree of luminosity. We discover the surface of the senile nucleus in relief. The embryonal segmentations are made visible both in youth and extreme age.

Of the vitreous body, we know but little as regards its exact structure, nor of the physiological remnants of the tunica vasculosa lentis, the vasa-hyaloidea propria (*Kölliker*), and the arteria hyaloidea.

The slitlamp discloses the living vitreous body in various manifestations of form, at times as a luminous wavering framework of folds, or the supporting structure is limited to a few scant filaments, membranes of definite form, fibres, or varying combinations of these components.

Of greater variety are the pictures of pathological changes in the vitreous body as revealed by the lamp: absorption of the supporting structure, senile and pathological hypertrophy, and opacification, deposits of crystals, blood, lymphocytes, pigment etc., are open to inspection. In addition we are enabled to *localize* exactly areas within the eye. We are given an "Optical section" of the living eye, and the location of areas in cornea, lens and vitreous can be as exactly fixed as in anatomical preparations or slides. Everyone who has become acquainted with this new method, consisting in the combination of the slitlamp with the corneal microscope, will agree that with it ophthalmology has entered upon a new stage of development. For clinical observations, especially in the field of eye-injuries, the slitlamp reveals new symptoms for the purpose of an early differential diagnosis, of momentous importance to the welfare of the patient, and to scientific ophthalmological research.

The slitlamp has attained its full practical value only through its combination with the Zeiss-Czapski binocular corneal microscope. The combination of the two instruments is to the credit of Henker, who has thus made microscopy with the illumination of Gullstrand possible.

By means of it, we can examine the anterior eyeball with a lineal magnification varying from 10 to 68 times. For more minute observations we have a magnification of 86 to 108 times at our disposal.

Very few authors have as yet adopted this new method of investigation, owing to the fact that the discovery of this light just preceded the outbreak of the world war. Consequently, its importance is but little known. Another reason may be that improvements in the apparatus are of very recent date. Its manipulations are not so easily acquired, in fact we may say that its successful application requires as much practice as the use of the ophthalmoscope.

It is the object of this atlas to in a measure remedy these deficiencies and furnish full instructions for methodical observation. To all of the aforesaid difficulties we must add that the high cost of the apparatus has also prevented the rapid adoption of this new method.

Koeppe recently described an apparatus by means of which he made the fundus and macular zone accessible to microscopy with the slitlamp. I have not yet had an opportunity to test its application.

B.

TECHNIC

In regard to the construction of the illuminating lamp and corneal microscope, we refer to the reports of *Gullstrand*[1])[128]) and the description furnished with the apparatus by the Firm of *C. Zeiss*.

The method is based on utilization to the limit of the principle of focal illumination. The advantages of the latter are a lateral illumination by means of an excessively bright sharply defined bundle of light, with a dark background.

By means of the slitlamp we take full advantage of these facts. The light of the Nernst-fibre is not thrown directly into the eye (of late also the nitrogen lamp*), but at first filtered through a diaphragm, provided with a narrow slit. This eliminates aberrant rays. Then after passing a diaphragm of suitable size the bundle of light is projected into the eye by means of a spherical lens.** *The densest focal part of the luminous shaft is projected onto the area which we wish to examine.* If we use a rectangular diaphragm, the bundle of light will also be rectangular, and for a short distance prismatic (Fig. 1). If however, the opening is *round*, the focal bundle of light will be approximately *cylindrical*. ("Cylindrical bundle" in contrast to slit-bundle, see below.)

Within the last two years I have introduced a modification of the Gullstrand apparatus[76]). Instead of focussing the Nernst fibre within the slit opening, I let the image fall on the *diaphragm opening of the illuminating lens.**** By this means a portion of the light otherwise lost is retained, and the fibre image is of greater luminosity. In addition the bundle of light gains in distinctiveness, and is more homogeneous. The present form of burner and length of its chamber does not allow of this modification. Either a burner with a longer base or a shortening of the chamber is necessary. (This latter change may be made by any skilled mechanic.) This modification (Köhlers method) was subsequently applied to the nitrogenlamp by *Henker*, later by Dr. *Streuli* (University Eye Clinic, Berne), and then by my assistant Dr. *W. Schnyder*. (The chamber furnished by the firm of Zeiss was correspondingly reduced in length.) The specific luminosity of the nitrogen fibre is greater, and its light is whiter than that of the Nernst fibre. However the fibre image is less homogeneous, and shows coloured lines and stripes. *These latter defects are practically eliminated by this new method of focussing the image. The less expensive, specifically more luminous, and more easily handled nitrogen lamp, is for this reason to be preferred to the Nernst fibre.* For certain purposes, for instance, in the determination of depth, a reduction of the size of the slit to $^1/_2$ mm or less, is absolutely necessary.

Aside from the fact that the examination is carried out in a dark room, a certain dark adaptation is necessary for more minute investigation of such structures, as for

---

* In many instances a substitution of a micro-arclamp is suggested. This has been especially mentioned in the discussion of the examination of the cornea and vitreous body.

** I have found a simple spherical lens sufficient for all ordinary purposes. For purposes of the "Method of the cylindrical bundle", as described below, an aspherical lens is however to be preferred.

*** As the latter is movable, it is best to choose a medial point of location when determining the focal point.

instance the vitreous. As a diaphragm[2]) for the illuminating lens, we use one of approximately 10 by 16 mm. (Supplied by the firm of *C. Zeiss*). The illuminating lens had to be somewhat flattened on one side, so as to avoid conflict of the axis of light with that of the microscope, when the two present a very acute angle to one another.

This slight change has made it possible for us to explore a much larger part of the posterior lens surface and vitreous body. Aberrant light can be excluded by a screen of black cloth or paper. Koeppe's tubular screen[3]), which allows of the insertion of coloured glass filters, is also useful.

The arrangement of the Zeiss binocular microscope is that devised by Henker[4]), with movable base[2])[76]), adjustable forward, backward, and from side to side, by rack and pinion. The latter is as much of an improvement compared to manual adjustment, as is the micrometer screw in a microscope in comparison with focussing by hand. A much greater delicacy of movement is obtained. Chin support and headrest are essential.

During the examination it is necessary for the patient to use the other eye in fixing an object, which later can be placed in various positions. For this fixation we use a light of very low candle power.

We use ocular No. 2, 4, 5 and 6, as objective No. F55, a2 und a3. The combination of ocular 6 and objective a3 gives a linear magnification of 108; with it however, the physiologic oscillation of the eyeball is sufficient to interfere somewhat with accurate observations. The greatest magnification ordinarily employed is one of 86 times (Oc. 5, Obj. a3). The one most commonly used is one of 24 times (Oc. 2, Obj. a2).

The binocular microscope is well adapted for investigating the *dimensions* of ocular structures. Formerly we rarely used the *measuring ocular* basing our observations of the size of objects on comparison by their contrast with the iris pigment border, or the diameter of blood vessels.

This inexact method leads to error and must therefore be discarded. For this reason we now use the *measuring ocular*[5])* and find it of great value, especially in lower powers of magnification (10 to 37 times).

We are now in the position to measure accurately: corneal opacities, infiltrates, vacuoles, cystic spaces, precipitates, nodules of the iris, the diameter of vessels, reflected images etc., and to observe their variations.

This result cannot be too highly appreciated in its relation to scientific research and value in clinical application**. The fixed dimensional value of a degree of displacement on the scale of the measuring ocular when in combination with a certain objective, can be ascertained by using the glass slide of a Zeiss blood corpuscle counting apparatus, fastened on white paper, as a guide, viewing it through the microscope in transmitted light. Investigation of the increase in size of objects as seen through the anterior media show a magnification of about $1^1/_{12}$ to $1^2/_{12}$ times in the zone external to the pupillary border, in eyes with normal corneal curvature and depth of anterior chamber, an amount relatively unimportant. In the posterior lens zone and immediately behind, the magnification is probably less than $1^1/_2$ times.

In obtaining these measurements we used a needle graduated at intervals 0,5 mm (constructed by James Jaquet, Basle). The needle with the visible graduation was

---

* *Stargardt*[115]) was the first to apply the ocular-micrometer to the binocular microscope.
** We can for instance, by measurement, keep control of variations in the size of infiltrates, vacuoles, fissures in the lens, iris tubercles, etc.

passed through the anterior chamber in the plane of the iris of freshly enucleated eyes, taking care that no aqueous fluid escaped.

In a similar manner we examined a needle with scale, when inserted through the vitreous body parallel to the equator at the posterior lens pole (Position ascertained by slitlamp.) Before and after these measurements the refraction of the cornea was ascertained. As the corneal epithelium suffers, it was necessary to moisten it with a Ringer's or similar solution just before making the observations. In some cases it was necessary to wipe off the defective epithelium in toto.

It would be of greater service if the slitlamp had a device, by means of which the angle of the axis of one of the eye pieces and that of the direction of light could be measured. The transporter or slide which we are compelled to use is correspondingly inaccurate. Were we able to gauge the angle, we could compare an object under the same conditions at different times.

Furthermore we could measure the depth of the anterior chamber, the thickness of the cornea, and especially that of the lens. This axial diameter of the lens as well as that of the nucleus in the same individual varies in age. It would be of interest to be able to measure it approximately.

Fig. 91c a very thick lens.
Fig. 91d an ordinary lens.

In the first case the swelling may have been caused by absorption of fluid (intumescence).

The ability to so compare lenses would be of value in the study of the genesis of senile cataract. We could for instance, establish comparative values by having the patient look into the shaft of light, determine the distance x micrometrically, using a constant angle* (Fig. 91b). The length of a, the axial diameter of the lens is therefore approximately $\dfrac{x}{\sin \gamma}$.

The deviations due to the corneal and lens surfaces are not considered. A series of measurements have proved that the figures given are useful as comparative values.

---

* The firm of Zeiss (Prof. Henker) at our suggestion, has applied a useful gauge for measuring the angle. A metal transporter rotating on its centre, with movable ruling, is attached to the vertical rotating axis of the binocular microscope. This arched transporter is attached to the illuminating arm of the slitlamp, which allows of a reading of the angle between the direction of illumination and the central axis of observation. It is accurate to within $2\frac{1}{2}°$.

C.

METHODS OF EXAMINATION

The slitlamp allows of four methods of examination*.

1. *Direct lateral (focal) illumination.*

2. *Transillumination* (diaphanoscopy). Examination in transparent light, as reflected by the iris, the lens etc.

3. *Direct lateral illumination* of *reflecting* (limiting) *surfaces.* This method permits examination in diffused light, more particularly in the light of the (circumscribed) *reflecting zones. (Circumscribed,* intensely luminous *areas* on the limiting surfaces of media, *in exact focus).*

4. *Indirect lateral illumination,* by means of which singly or multibly reflected light is utilized at the margins of illuminated areas.

Method No. 1, the one originally employed, is the most important. In combination with it and stereoscopic inspection we obtain plastic pictures in their natural form and colour. In this last respect method No. 1 is superior to method 2 and 3.

The second method (transillumination) is used principally for the study of surfaces which are apparently covered by a dewlike deposit (changes of a dewlike nature), also for observing the circulation of the blood and for transilluminating the iris etc.

Method No. 3 was but recently introduced in slitlamp-microscopy. Thanks to this method we are now able to observe the living endothelium of the cornea. Every individual endothelial cell becomes visible.

In the lens the epithelium and the posterior lens surface, its graining (shagreen) and the changes in the latter come into view.

This method and method No. 1 are both of value for the purpose of localization (determination of depth).

The one of least practical value and importance is method No. 4. It seems to have been little used. It is however (as in ophthalmoscopy) often of value, for instance in observing dewlike changes of the cornea and certain lens conditions, especially the vacuole formation in the subcapsular areas.

We will now consider the various methods of illumination, with regard to their applicability and practical importance and examine them in conjunction with the illustrations of this atlas.

---

* All of these methods of illumination have been practiced when using ordinary focal illumination. With the slitlamp they can be more easily and accurately differentiated from one another.

1. EXAMINATION IN FOCAL ILLUMINATION DETERMINATION OF DEPTH

Helmholtz[6]) observed more than sixty years ago an *internal reflection* in the cornea and lens in intense focal illumination. The cornea and lens, as well as most organic and inorganic refracting media, are to a slight degree opaque, not absolutely transparent. The various individual tissue components differ in their refractive indices, they are heterogeneous.

Recent observers have confirmed this, *Gullstrand* among others.

According to *Stokes*[7]) we can divide the light diffused by an illuminating media, into reflected (dispersed), and fluorescent light. He speaks of a "real" and "false" inner dispersion, meaning fluorescence by the first named.

The physicist *Spring*[8]) has proved that there is no "optically empty" fluid in nature. Every media is defective in the respect that some of its component particles are media of slightly varying refractive indices. He has however produced an optically empty fluid in the following manner: An electric current was passed through a U-shaped tube filled with water and containing quartz-powder in suspension. The current caused the particles to collect about the cathode, whereas the liquid around the anode remained perfectly clear.

A ray of light passing through it was rendered invisible when viewed from the side, proving that the fluid was in reality "optically empty". (A. Winkelmann, Handbuch der Physik, 2nd Edition 1906, Optik, P. 788.)

Among the transparent media, the cornea, lens and retina, are most opaque and fluorescent. The aqueous is nearly "optically empty", whereas the vitreous body varies in individuals, as demonstrated by the slitlamp. The opalescence of cornea, lens and vitreous body increases with age.

By excluding the rays producing fluorescence with a filter of yellow glass, as has been done by the author[9] [10]), we can easily show that the light dispersed by lens, cornea and vitreous body in focal illumination is fluorescent only in the smallest degree*.

The reflected light is but slightly reduced in quantity by the exclusion of the rays above mentioned.

(The fluorescent light alone can *only* be demonstrated by the projection of an arc-light through an *Uvialglass* [Schott & Gen., Jena], or by filtering it through a solution of oxide of ammoniated copper, or still better by pure ultra-violet light [*Lehmann's* ultraviolet filter].[10]) These two phenomena, opalescence and fluorescence, are not peculiar or unusual, they are characteristic of all organic and of almost all inorganic bodies, to a certain degree. (In ophthalmology these rays have attained a certain notoriety, because some authors were inclined to ascribe the origin of senile cataract, erythropsia, retinitis, retinal atrophy etc., to their irritating influence.) Opalescence and fluorescence are best observed in a dimly lighted room. The more circumscribed and intensely luminous the shaft of light, and the darker the immediate surroundings, by contrast, the more vivid do they appear.

---

* Respecting the retina, there have been no known observations in this regard. However, superficial retinal folds, after the 35th—40th years, show a dull surface when observed with the red-free light, while in youth they show a greater reflection, as has been observed by the author in very many cases.

Where the bundle of light of the slitlamp passes through cornea and lens, for instance, Fig. 15 shows an optical section through the cornea, Fig. 127 through the lens, these tissues appear opaque. The illuminated zones are translucent only, no longer transparent. Areas of *increased* refraction are more conspicuous. We can now see lens opacities of slight degree, invisible with ordinary focal illumination. If we let the light enter the eye in the antero-posterior (respectively meridional) direction, we obtain (sagittal) optical sections of the tissues (Fig. 1—cornea). The anterior chamber, between the luminous lens and cornea is now relatively "optically empty", therefore dark. By changing the position of the arm of the lamp one can, so to say, gauge the depth of any part the anterior chamber. The presence of exudates and cells is betrayed by the luminous shaft becoming visible in the anterior chamber. In this manner we are able to detect the most minute diffuse exudation in the aqueous, and distinguish it from diffuse corneal opacities. Such differentiation was heretofore impossible.

In some cases, for instance, after operations for cataract or glaucoma, the anterior chamber is so shallow that we are in doubt as to whether it may exist. In regard to this the slitlamp can alone give us reliable information. I have in this manner repeatedly found that aqueous was present only in the pupillary area, for this zone was "optically empty". Its antero-posterior diameter corresponded in depth to the thickness of the iris. In a similar manner we can determine the absence or presence of a space between the posterior corneal surface and the apex of an iris tubercle or "iris bombe", or we may, in cases of peripherically situated anterior synechiae, exactly determine their boundaries of adhesion. Following an evacuation of the anterior chamber, the latter may be obliterated in the area of the thickest part of the iris, that is at the circulus minor, while peripherally a slight separation may be demonstrable with the slitlamp.

Where the tissues are in contact with the posterior corneal surface they replace the refractile area between cornea and aqueous (Fig. 1a, efgh). The same perfection of observation can be attained in determining the presence of a possible space between pupillary border and lens. (In cases of subluxatio lentis, shrinking cataract, or forward displacement of the iris.) Very small separations are easy of detection in this manner.

The *determination of depth*[15])[100])[103]), is an important advance in clinical diagnosis achieved by the use of the slitlamp. Depth can be determined in two ways*:

1. By means of the practical and simple method of binocular stereoscopic observation. This method is quite easy for coarse differentiation, inefficient however for determining the depth of very shallow areas.

2. By means of the observation of the *linear optical section*. Other authors who have investigated with the slitlamp have not as jet mentioned this exact line of demarcation, they speak of an optical section (Optischen Schnitt), meaning thereby the whole of the area rendered opaque by the bundle of light. We will show that for the *exact* determination of depth a distinct surface must be utilized.

We will explain this by taking the cornea as an example (Fig. 1a). *For the purpose of determining depth the size of the slit must be reduced to a minimum of about 0,5 mm. The axes of observation and illumination must not present too acute an angle*

---

* Parallax, perspective displacement, and shadows in determining depth are of minor importance in slitlamp investigation, of greater importance is microscopical focussing. (Compare the chapter relating to dewlike changes—Betauung. Regarding measurement of depth by reflecting zones, see Method 3.)

to one another, at least 60—80°. If we project the densest part of the luminous shaft diagonally through the cornea, from the temporal side, using a medium sized slit, the illuminated tissue zone will have an approximately prismatic form (Fig. 1a). Anteriorly this prism is bounded by the *entering surface of the light abcd*, on the side of the aqueous chamber by the *surface of exit efgh*. We can easily distinguish the three edges *ac, bd, fh*, less distinctly the fourth edge *eg*.

Most important, though for the beginner somewhat difficult, is the determination of *edge bd* which bounds the entering surface on the nasal side. A drop of fluorescein will decidedly simplify the demonstration, and edge *bd* will now appear sharply defined, in the normal as well as in the pathologic cornea (Fig. 15). This edge bounds the important field *bdfh* (dotted line Fig. 1a) which represents a *linear optical section* through the cornea, now easy of observation. The observer must alternately focus the edges *bd* and *fh*.

For the determination of depth, as has been stated, binocular observation is sufficient. (Exact estimates cannot be made by it, in fact, errors in observation may often occur.) On the other hand the *linear optical section bdfh* is of valuable aid in *localizing opacities.*

We let the luminous bundle pass over the area in question, and can localize any change (opacity), by its appearance and disappearance in the linear optical section, in the same manner as we can in a section of tissue under the microscope. If we wish, for instance, to determine the location of an opacity in the deep parenchyma, we first of all direct the luminous bundle toward the temporal side of the opacity. (Taking it for granted that the source of light is from the temporal side of the eye under observation.) We then approach the opacity until it appears in the optical section. Thus we determine its position. Naturally we can in an inverse manner let the opacity leave the bundle of light and enter the optical section. The position of the opacity in the latter *bdfh* is then given by the point at which it leaves the bundle of light. We may now microscopically determine the relation of the opacity to *bd* and *fh*, which will localize it within the superficial, medial or deep parenchyma, as the case may be. If it first appears in *bd* it would be situated in the anterior corneal surface. If first seen in *fh* it would be localized as being on the posterior corneal surface.

For the purpose of determining the edges *bd* and *fh*, the microscope and the bundle of light must be individually focussed on them. Up to the present this optical section and the edge *bd* bounding it anteriorly have not been described. The importance of the luminous shaft or bundle in its entirety only, have been described as an "optical section". It is evident that for exact localization, the described optical section only can be utilized and that it must be absolutely differentiated from the surface of entry *(abcd).*

(Monocular observation of the latter is sufficient, however binocular determination can be made with more ease.)

The surface of exit *efgh*, the posterior surface of the prism of light, reflects decidedly in cases of iridocyclitic deposits, and especially after parenchymatous keratitis. It can appear just as clear or clearer in diffused light as the surface of entry *abcd*. Its temporal edge *eg* is easily distinguished in these cases. Deposits naturally appear less distinct the nearer they are situated to edge *eg*. They are most distinct in the region of *fh*, as here they are not, as in the case of *eg*, covered by an illuminated layer (Fig. 20).

The above mentioned *reduction in size of the bundle* attained by narrowing the slit is especially valuable in determining a thinning of the cornea, partial obliteration of the anterior chamber, and localizing within the lens substance. —(Diameter of focal bundle about 0.05 mm.)

During the past year the author has originated another method of determining depth by first using the slit bundle of light in the manner indicated and then, for the purpose of controlling detail, applying the *method of utilizing the cylindrical bundle of light*. On substituting for the slit diaphragm one of the small openings adjacent to it* (preferably the one of 1 mm diameter), the focal bundle area will become nearly cylindrical. The shape of the bundle may now be compared to the area created by a drill in passing through a solid object. With this illumination directed at an angle one may easily observe the steplike arrangement of layers, as if they had been punched out ("cylindrical bundle" in contrast to "slit bundle").

Nitrogen and *archight* (when the image of the fibre is focussed on the illuminating lens as described on page 3) are also to be preferred to the Nernst light. The angle between observer and the direction of illumination must be chosen relatively large (60—70°). By our observations comparative freedom of colouration and distinctiveness of the bundle are attained by the application of the principles of aplanatic and achromatic optics**. Fig. 1b shows the cylindrical bundle of light as it passes through the cornea. Anteriorly and posteriorly it is sharply bounded by circular white areas about 0,25 mm in diameter, $V=$ anterior corneal surface, $H=$ posterior corneal surface. By narrowing the slit, the cylinder of light may be flattened from side to side.

This method has proved efficient for purposes of refinement in localization. It should be used in conjunction with the first described method, which latter allows of a better orientation, for the control of detail. For example: for an exact determination of the depth of vessels and nerve fibres, of *striped* and *flat* opacities within the various media, of variations in the thickness of the cornea, and especially for the diagnosing of *partial* or *total* obliteration of the anterior chamber. (For further detail regarding the determination of depth, see the respective chapter.)

When examining in diffuse light, the angle we generally use between the medial axis of observation and the direction of the light is one of 40°. The patient looks straight ahead. Changes in the media appear in their natural colour. Those of the lens and vitreous are correspondingly influenced by the yellow colour of the lens. (More in age than in youth.)

Precipitates appear gray to whitish-gray, and if brown the lustre is in proportion to the amount of pigment they may contain. The smallest conglomeration of cells, even individual cells are discernable with high magnification (lymphocytes, red blood cells), though they are less sharply defined, than with method 3.

Opacities of the lens, if very thin and flat, appear blue to bluish-green, for instance the wreath-shaped cataract often shows shades of this colour. (Also cataracta coerulea, viridis etc.) In transillumination however they are brown to brownish-yellow. Regarding the origin of this green tint compare[9]) of the author. With the aid of an emulsion the formation of these "physical" colours can be instructively demonstrated[9]).

Opacities in the lens of greater density appear white. (In transillumination they are black.) Red blood cells are a luminous light yellow, not as *Koeppe* states a brick-red (to avoid error one can make comparison by applying a thin smear of blood to a coverglass), pigment particles appear brick-red to brownish-red. (Compare "Pigment particles of the supporting structure of the vitreous", Fig. 336, 353).

----

* These openings are added by the makers of the lamp for a purpose other than the one here mentioned.

** We have installed an achromatic-aplanatic collector in the lamp chamber and chosen an achromatic illuminating lens. The diaphragm of the latter is round and measures 10 mm (exact centreing is of great importance!) Thereby we have practically eliminated the annoying coloured border and halo of light otherwise present in the cylindrical bundle. The slit bundle can also thus be made more homogeneous in character, to a degree heretofore impossible.

Pigment particles, the diameters of which are less than one micron, are visible by reflection. Such pigment dust can be found in the vitreous body and also on the posterior corneal surface after trauma. (The elements of the retinal iris pigment may be demonstrated by macerating a piece of freshly excised iris on a glass slide.) Such elements of pigment are not visible in transillumination with the magnification employed. In the direct focal illumination with the slitlamp they appear as luminous spots within darker surroundings, comparable to the dust made luminous by rays of sunlight in a dark room.

The slitlamp cannot aid us in determining the various sizes of these small elements. We must beware of attempting to judge their real dimensions by their apparent size; this would lead to error, as has been experimentally proven[11]). The more heterogeneous the medium, the more intense its reflection, when illuminated with the slitlamp.

In parenchymatous keratitis, this reflection is so pronounced that details of deeper situated parts may be veiled by the light diffused by the more superficial ones. We have shown by means of Fig. 1a and Fig. 20 that opacities of the posterior corneal surface are less distinct, the further they are situated from edge fh.

In parenchymatous keratitis, the heterogeneous character of the changes in the cornea around newly formed blood vessels may be so marked that the latter may be quite invisible in diffuse focal light, while with transillumination they may still be seen, as well as the blood corpuscles in them, as they roll along (compare Fig. 19). The first observer who saw the circulation of the blood in the conjunctiva was *Adolf Coccius*[119]) (1852), *Donders*[120]), *Bajardi*[136])[137]), *G. Schleich*[116]), *K. Augstein*[118]) *K. Stargardt*[115]) and others.

For similar reasons the *vascular loops at the limbus* in normal eyes are visible in transillumination, and veiled or hazy in direct light. On the contrary, homogenic media offer more favourable conditions for observation in direct than in transilluminating light. Thus we can easily discern in direct illumination the characteristic striations of keratoconus (Fig. 59—60). In almost normal corneas the corneal nerve fibres appear especially luminous (Fig. 48, 49, 61). Ruptures of Descemet's membrane may become visible.

In the normal lens, focal illumination will show the lamination of the lens, which latter we are unable to observe under any other conditions (compare Fig. 1—27). The embryonal sutures and their fibrillations appear, as also the numerous remnants of the anterior and posterior vascular membrane (Fig. 141—147 etc.). We see the supporting structure of the vitreous body, its normal and pathologic opacities, changes in density, and the remnants of embryonic vessels. To obtain a more general view for the purpose of orientation, one should substitute a broader bundle of light for the focal one. By moving the illuminating lens the field may at will be rapidly enlarged. The homogeneous character of the field is dependent on the form of the image of the fibre. If the image of the fibre is projected in the above described manner onto the illuminating lens, the extra-focal bundle areas will be inhomogeneous (permeated by coloured zones). If however the image of the bundle (according to the original description) is focussed within the slit, the annoying colours will naturally be found only in the focal area, while the extra-focal bundle areas will remain relatively homogeneous.

2. OBSERVATION BY TRANSILLUMINATION

Transillumination is accomplished by directing the bundle of light derived from the slitlamp onto more or less opaque tissues (iris, exudates in the pupillary area, cataract), from which latter it is reflected through the tissues in front of them.

Transillumination is therefore chiefly utilized in examining the *cornea*. It can however be used to advantage in examining the iris, and more especially the lens.

On the cornea it brings to view *dewdrop-like changes* or *deposits* on the *epithelium* and *endothelium**. It was first observed by *J. Staehli* on the endothelium of the posterior corneal surface, and its nature has not as jet been satisfactorily explained. Recently *Koeppe*[12]) endeavoured to explain this phenomenon.

These "dewlike changes" can only be observed in transillumination, in the light reflected by the iris or a hazy pupillary zone. Regarding its nature various suppositions have been made, which latter we shall not now discuss. It is certain that *dewlike changes* of the *endothelium* have been confounded with those of the *anterior epithelium* of the *cornea* (for diagnosis see below). Fig. 19 shows dewlike changes as seen with a magnification of 68 in a case of tuberculous iridocyclitis.

By focussing the posterior reflecting surface of the cornea and letting the light pass over areas of the endothelium showing this dewlike change, we can show two causes for the origin of the latter. *In the first place,* a fact heretofore unknown, the endothelium in transmitted light may assume the appearance of very minute droplets.

As in cases of oedema of the epithelium, endothelial cells also apparently become visible when their refractive index is changed by absorption of fluid. A uniformly delicate dewlike change results, visible only under high magnification.

Secondly, participating in the change are also cells deposited in the endothelium, which latter are usually larger than those of the endothelium. They often appear in groups and show an irregular form. Occasionally precipitates of cell masses, small particles and delicate fibres may appear transparent in transillumination. They can be distinguished from other droplets by a slightly reduced transparency. Larger precipitates show as shadows, especially if they contain pigment. In this connection note the transilluminated precipitates of Fig. 20, to the left, which show in part a concentric striation. In the right part of the illustration one sees the precipitate in *focal illumination,* they appear white, those containing pigment however are brown.

A comparison of Fig. 20 and 23 illustrates the manner in which three different methods of illumination may show precipitates. The most distinct illustration is the one obtained by the method of observing reflecting zones (see below, and Fig. 23). It also explains the relation of deposits and dewlike changes to the endothelium.

We can therefore differentiate between two types of droplets in characteristic cases of dewlike changes of the posterior corneal surface under high magnification (33—86 times).

1. The uniform fine droplets with which the epithelium is carpeted.

2. The round oblong or often clubshaped, slightly less transparent droplets which, like impurities, are irregularly scattered among the others. With reduced magnification (10—24 times), the latter only are visible.

Staehli has regarded these dewlike pearls as minute corpuscular deposits, and observed all transitions, from the finest dots to whole precipitates.

---

* *Translators note:* "Taubeschlag" is the name introduced by *Staehli* for the phenomenon which *Vogt* now calls "*Betauung*".

The elements of real deposits appear black in the reflecting surface (Fig. 23).

When observing the reflecting zone (Page 15) by this method, the endothelium appears normal in ·size, although the *margins of the individual cells* are *much less distinct than normal, sometimes unrecognizable* (see Fig. 24).

This picture presents the reflecting zone of the posterior cornea in a man, aged 35, who has a bilateral subacute iridocyclitis, and secondary glaucoma. In Fig. 23, the endothelium shows a dim or blurred appearance; in this case the latter condition is more pronounced. The endothelial borders are only visible in the most superficial corneal areas, even there they appear indistinct. The endothelium has assumed a darker colour, so that the lymphocytes deposited on it are not sharply contrasted. Endothelial dewlike changes are present. Tension at the time R. 28, L. 30 mm Hg. (after iridectomy).

In the course of examination, we occasionally see the layer of intact endothelium in the form of fine droplets, under the precipitates.

(As *E. Fuchs* has shown, the endothelium may be intact under precipitates, though sometimes it has in part disappeared.)

The fact that the frequent dewlike changes of the epithelium resemble those of the endothelium, may lead to the one being mistaken for the other. Both are the result of oedema of epithelial cells, therefore the similarity of the clinical picture. The differential diagnosis is not necessarily simple. If there are opacities in the cornea the latter may be of aid in localizing the areas of dewlike changes. If the latter are in the epithelium they will cover the opacities, if in the endothelium, they will disappear behind them.

If the cornea is clear, the bundle of light will be of no aid in localizing dewlike changes; because we are observing by transillumination, not in focal light. Anterior to the section *bfdh* Fig. 1 a, the epithelial droplets are invisible, and cannot for that reason be traced to the edge *bd* (see Fig. 15 and Fig. 20). Next to the aid given by the presence of corneal opacities in localizing these changes, the method of high magnification is of service, the point of focus determining whether we are at the epithelial or endothelial layer. The decision is made with ease, by observing the bundle of light (Fig. 15). Of further aid may be precipitates on the posterior corneal surface. On the anterior surface we may observe corpuscular elements of the tears, set in motion by the blinking of the lids.

The dewlike changes of the epithelium seem like a carpet of vacuoles of various sizes (Fig. 15). This is caused by an oedema of the epithelium. The anterior corneal reflecting surface may show globular elevations, which latter create numerous small reflexes (Fig. 14). The prominences are not always present in dewlike changes of the epithelium. Furthermore they temporally disappear on winking, while the dewlike changes of the epithelium *remain unaltered*. The dewlike changes caused by the oedema of the epithelium are less uniform than those of the endothelium. The globules or pearls of the epithelium are of various sizes (Fig. 15—16), while those of the endothelium are like a carpet of uniformly small droplets, therefore not as easy of recognition.

A further typic point of differentiation is furnished by the before mentioned, varied sized, large, scatteringly *deposited droplets* and delicate fibres, which are lymphocytes, or possibly amorphous clumps and fibres, and therefore found only on the posterior surface of the cornea (Fig. 20).

A comparison of Fig. 20 with Fig. 23 will show that the grouping of the club and dumb-bell shaped dew-drops is characterized by a certain concentration or apposition, because they owe their origin to the junction or confluence of individual lymphocytes or albuminous clumps. Thus chainlike strips or clusters of droplets originate. (These cells only present sharp outlines, when the microscope is focussed on the reflecting zone, and not when focussed on the dewlike changes.) At times, in cases of severe iridocyclitis, especially so in glaucoma, in parenchymatous keratitis, and after perforation, we may observe dewlike changes of both the epithelial and endothelial layers of the cornea.

In all eyes with *normal corneae*, I have noticed a delicate dewlike change located in the epithelium near and in the limbus, with indirect illumination. This phenomenon is therefore found under *normal conditions*. The epithelium appears as if composed of fine droplets in transilluminated light. They are especially distinct between the fine vascular loops of the limbus. To observe this it is necessary to direct the illumination from the opposite side into the angle of the anterior chamber.

This appearance noted in the epithelium normally, may be utilized in localizing other areas; it also places a (pathologic) value on similar changes of the endothelium. It may be mentioned that we have noted dewlike changes on the surface of old corneal maculae, often visible decades after the occurrence of an infiltration.

The fact that we can recognize *vascularization of the cornea* more easily by transillumination than by focal light has been emphasized and explained (page 11, see Fig. 9a—57).

For the same reason we can rarely see the circulation of the blood with focal illumination, it is much easier of observation by means of transillumination. It is not difficult to observe the circulation in the vessels at the limbus in normal eyes.

For the iris, transillumination is of value in cases of atrophy of the pupillary border, and of its retinal pigment layer. In age the (retinal) pigment usually disappears, leaving a membranous translucid sheath (Fig. 300, 302), which latter is visible in focal light as well as in transillumination. This sheath is not always present, though the pupillary border is free of pigment, the latter having been scattered onto the iris, or into the aqueous. (Such distribution of the pigment takes place, in a greater or less degree with regularity in advanced age, and need not be, as was stated, the sign of a tendency toward increased intraocular tension.) In these cases the whole of the sphincter area of the iris may be more or less translucent (Fig. 305). In transmitted light, this zone resembles a thin translucent membrane, whereas in focal light the tissue may appear quite normal. This acquired translucency is due to the disappearance of the retinal iris pigment in the pupillary area.

Suggillation of the iris (ecchymosis after contusio bulbi) is visible only by transillumination. The blood-stained area shows a distinctive red, when the bundle of light is reflected just *to its border*. In direct illumination the ecchymotic area shows no blood or colouring.

Atrophic areas, holes in the iris following atrophy, perforations, iridodialysis etc. appear luminous if we direct the light of the slitlamp into the pupil. We see these areas in the light reflected by the media posterior to the iris (compare also the method of diascleral transillumination, *Ruebel*, and others, and the transillumination through the pupil, according to *Staehli*).

In the lens, transillumination is especially important in determining the presence of the anterior subcapsular vacuole layer, which has been described elsewhere. The droplets are probably within or under the epithelium. Similar vacuoles are also seen in the cystoid spaces. These latter are caused by the separation of the fibres and lamellae of the lens. They are best visible in transillumination. (Indirect light from the side.) In focal light they appear white (Fig. 208). In the lens as well as in the cornea, transillumination often reveals deposited crystals of cholesterin, invisible in focal illumination, because they are veiled by opacities. Finally, the vessels of the conjunctiva, filtration cysts (cystoid cicatrices) and other changes are better seen in transillumination, than is possible in focal light. In addition to the fluid seen in filtration cysts we may observe in them pigment dust (*Erggelet*[102]) and debris from the interior of the eyeball, also incarcerated fragments of uvea, lens capsule and other tissues.

3. EXAMINATION OF REFLECTED IMAGES AND REFLECTING ZONES

A) IN GENERAL

Up to the present investigators have made no distinction between optical effects caused by reflecting (limiting) surfaces *of the eye media* and the results of *diffuse reflection of the tissues*. In fact, the question of such a distinction never arose. We will here explain the systematic separation of these two optical effects, as the results so obtained are of practical importance.

Direct focal illumination as has been described, is generally diffused light, that is, the different tissues reflect the light into various directions (see above). *On the surfaces of the limiting optical media, we see in addition to diffused light, definite reflected images.*

We will now discuss the latter, and the reflecting surfaces producing them. Many observers have, without doubt, considered these reflected images a nuisance, and endeavoured to eliminate them. *Hess*[45]) had however, long before the introduction of the slitlamp, discovered the anterior graining (shagreen) of the lens, and in so doing observed the reflecting zone of the anterior lens surface. The *anterior lens zone* as such shows no graining, the latter is only visible when the area is in *exact focus*. It is never seen except at the lenticular reflecting surface, which latter produces it. *Staehli* had examined the anterior corneal surface "in the reflex". *Koeppe*[3]) reports having seen detail of the posterior surface of the lens "in the reflex". This differentiation of reflected image and reflecting zone has hitherto neither been mentioned by other observers, nor applied to any of the various eye media. For practical purposes, we can observe the differentiation of reflected image and reflecting zones of the corneal anterior and posterior, or the anterior and posterior surface of the lens, with the aid of a glass lens. If we direct the corneal microscope toward a concave reflecting surface, we can find the reflecting zone only if we focus onto this surface or its irregularities.

With the slitlamp and corneal microscope we observe as follows: We direct the microscope onto the reflected image which we wish to study. The diaphragm of the illuminating lens we use produces reflecting images, vertically quadrilateral in form. The focussing of these images, depending on which surface is in question, is more or less difficult. The easiest to focus are the luminous clearly outlined images of the anterior corneal and posterior lens zones. They can be easily fixed because of their luminosity, which latter makes them so disturbing in ophthalmoscopic observation.

The image at the *anterior lens zone*, owing to its position and indistinct outlines, is somewhat more difficult to focus and observe. A still greater difficulty is noted when examining the image of the *posterior corneal surface*. It is necessary to so remove the latter image that it will not conflict with that of the more luminous one of the anterior surface, in a manner to be indicated.

If we have (with a magnification of about 24 times), directed our attention to a certain surface reflection, and brought the same into the centre of the field, we must so regulate the tube, as to bring the reflected zone into an exact focus, in order to see the detail on it clearly defined.

(For instance, on the anterior corneal surface we must see the corpuscular elements of the tears, and on the anterior and posterior lens surface the graining [shagreen].)

The reflected image will be somewhat less distinct, but the reflecting zone can be studied in all of its luminous detail. The *reflecting zone* is now in focus[14]).

By altering the angle of observation, or the angle of incidence, or both simultaneously, we can let the luminous reflecting zone *wander* to any part of the respective reflecting surface under observation. If, for instance, we are focussing the reflecting zone of the posterior lens pole and wish to examine the capsule below this area, we direct the patient to look downward; if we wish to let the illuminated zone wander nasalward, we direct the patient to look nasalward, etc. To explore surfaces that are *convex* anteriorly, such as the anterior and posterial surface of the cornea, and the anterior surface of the lens, the patient must be directed to look in a direction opposite to that in which we wish to examine.

With low and medium magnification the reflected zone in focus occupies only a part of the field under illumination. The image at the anterior surface of the lens is the largest, while that at the posterior surface corresponding to its radical curvature, is smaller.

Examine for instance, the small (slightly yellow) image on the posterior corneal surface (Fig. 2) and compare it with that of the anterior surface. Their size can be measured accurately and easily with the ocular micrometer.

The area surrounding the reflecting zone is seen in a very much reduced diffuse light, owing to the irregularity, fluorescence and opalescence of the media (see Fig. 3, D, D' = diffused light, Sp, Sp' = reflecting zone).

Impurities (defects and irregularities) in the reflecting zone show dark to black on a luminous background, while contrarily in the surrounding area, owing to the diffusion of light, they are luminous amidst darker surroundings.

The margins of the reflecting zones (Fig. 2) in comparison to those of the reflected images are somewhat indistinct and the corners seem rounded (Fig. 3). Also the zones are larger than their corresponding images.

The opacities and irregularities of a reflecting zone, which appear black on a luminous background, are comparatively speaking, like deposits or defects in the mercury film of a mirror. Consequently we can observe *changes* in the *reflecting zone*, *so minute* as to be invisible in diffuse light.

One must not underestimate the value of the reflecting zones in *localizing* changes in the immediate vicinity of the reflecting (limiting) surfaces.

If the reflecting zone is intact, the changes will not involve it, and therefore must be situated behind it. If they are in contact with it (for instance in the case of precipitates of the posterior cornea or lens surface), the contact area will create irregularities in the reflection. This is the case unless the changes present a similar difference in their index of refraction compared to that of the endothelium, as the latter does to the refractive index of the aqueous.

If such changes are visible in the reflecting zone, as fixed dark areas, it proves they are within the surface, or *anterior* to it. What is therefore the practical importance of the reflecting zones? We shall discuss the reflecting zones of the anterior and posterior corneal surface *(B)*, next those of the two surfaces of the lens *(C)*, and lastly the linear reflexes caused by folds in these reflecting limiting surfaces *(D)*. (For more detailed description we must refer the reader to our other publications.)

B) REFLECTING ZONES OF THE CORNEA

Observation of the anterior reflecting surface of the cornea is of less practical importance than of the posterior. Oedema of the epithelium (stippling) shows numerous delicate protuberances and irregularities. The margin of the zone is escalloped, and projects luminous plaques into the diffusely illuminated surroundings (Fig. 14) (the protuberances disappear for a time after lid movement).

The most *delicate irregularities* of the *cornea*, in no *other manner visible, can be seen*, when the *reflecting zone* is *in focus*. We also see the various interfering (dispersed) colours, caused by the layer of fluid on this surface.

Corpuscular elements of this fluid are visible as dark spots, often similar to the entoptic image. (Positive scotoma.)

A fine mosaic-like surfacing can be seen in certain areas, and at first glance one may take these to represent the borders of the individual cells of the superficial epithelial layer*. A greater magnification identifies them as corpuscular elements in the fluid layer, and movable with it.

Of much greater interest is the study of the *posterior corneal surface*, for on it we can see the *living endothelium*[15]), *sharply outlined* in the *reflecting zone* (see Fig. 3, *Sp'*). We are observing the *endothelial surface* adjacent to the aqueous.

The cells are hexagonal in shape, nearly all of uniform size, and form a honeycomb-like mosaic. Its colour is faint yellow with the Nernst light (compare the colour of the posterior corneal image Fig. 2).

The cell margins show as clearly defined dark lines. Nuclei are not visible.

Under a magnification of 24 times the cells are just visible. They are clearly seen with one of 37 to 68 times.

As Fig. 3 shows, the endothelial zone, that is the definitely outlined and focussed reflecting zone, is situated in the strip *D'* (surface *efgh*, Fig. 1a) which is created by the bundle of light on the posterior corneal surface by diffuse reflection.

In age the endothelial cells are less sharply defined than in youth, probably because they are veiled by opacities and the arcus senilis. (Gerontoxon.) Instead of the regular honeycomb-like pattern, a more amorphous granular surface, with indistinct cell outlines, is often found in otherwise normal eyes in old age (for instance Fig. 4 shows prominences)**.

The visibility of the corneal endothelium is of theoretical as well as of practical importance. Minute changes in it, oedema, defects etc., are made visible by the new method which cannot be observed by any other clinical means. In cases of precipitates we see every individual added cell, on the surface of, or among the endo-

---

* *Staehli* observed this with Azo-light illumination and presumed the marking to represent the epithelial cell borders (Klin. Mo. f. Aughk. 54, 686).

** *Salzmann*[16]) *anatomically* accentuated this blurring at the periphery in the areas of *Henle's* wart-like protusions.

thelial cells. They appear black on a luminous surface, as may be expected, for they present an interruption and addition to the reflecting surface in focus. (Fig. 23 shows an area of the field of precipitates in a case of chronic cyclitis.) We find these cells in great numbers in cases of iridocyclitis, when they are otherwise invisible with the ordinary illumination of the slitlamp. Similar to the cells we can also see the precipitates, black on a luminous background (Fig. 23). The precipitates are often surrounded by a wreath-like border of more luminous endothelial cells*.

Individual cells are often 'arrayed in chain-like strips or in clusters, reminding of certain bacteria (Fig. 23).

In cases of dewlike changes of the posterior corneal surface (see above) the cell borders are *less distinct*, in fact they may be invisible (see "dewlike changes" of the endothelium).

The shape of the normal endothelial zone of normal corneas when using a vertical slit, is that of a vertical oval (Fig. 3). It is irregular in irregular astigmatism, and especially so in cases of keratoconus and in scars within the deep parenchyma. One also learns that irregular astigmatism of the posterior corneal surface is *quite common*, and may be of a very high degree. Variations in the curvature of the posterior corneal surface, presuming the cornea is transparent, are manifested in the reflecting zone, as for instance: folds in Descemet's membrane, *ringshaped thickenings*, such as occur in cases of post-traumatic circular corneal opacification (Fig. 53—54), *circumscribed thickening* of the cornea in discform keratitis, *perforation scars*, etc.

These irregularities of the posterior corneal surface give rise to reflections, in the *reflecting zones* of which latter the *endothelium is discernible.*

For instance Fig. 27 shows the endothelial reflex with its delicate mosaic figuration near a perforating scar. (Particle of iron perforating cornea 3 years ago.) By observing the endothelial surface in an oblique direction, the cells appear shortened in their perspective. Finally we must mention that similar real or apparent circumscribed irregularities may be found on the posterior surfaces of normal cornea.

In all normal eyes I note a wavelike irregularity of the mosaic pattern of the endothelium; if this is not due to errors in observing the refraction of the light, it may be caused by irregularities in the curvature of the endothelial surface (see Fig. 3 and Fig. 5).

These apparent or real irregularities are more distinctly seen at the border of the zone. They are also more pronounced in age.

In correspondence with these clinical findings *Greeff*, in his "Pathological anatomy of the eye" [17]), mentions that Descemet's membrane is not perfectly smooth, as is the lens capsule, but shows delicate wave-like irregularities.

As Fig. 6 shows, the endothelium in age may take on an amorphous appearance. The cell borders are less distinct and the irregularities more numerous. A reduced transparency of the cornea may also aid in producing this picture. Exceptionally I have found the endothelial cell borders well defined in individuals who were in their sixth and seventh decade of life. In older individuals I have found circular elevations projecting into the aqueous (Fig. 5—6 etc.). These appear black, due to the changed angle of incidence. At their summits they show the endothelium, hence they can be differentiated from cellular deposits. (In the case of Fig. 29, non-progressing Kerato-

---

* By observing precipitates that are situated at the border of the endothelial field, we came to the conclusion that this wreath-like area of increased luminosity is due to an alteration in the surface curvature surrounding the precipitates (see Fig. 23).

conus, the endothelium appeared somewhat granular.) These prominences may be scattered or grouped. Their diameter varies from 20 to 100 microns. They correspond to the wart-like protrusions found most numerous in age, in the periphery of Descemet's membrane, which were first mentioned by *Hassal*[18]) and later more fully described by *Henle*[19]). These "warts" may be found in all adults.

Our ability to so carefully observe the endothelium will in the future enable us to diagnose ablations and detachments of it. Fig. 30 shows sharply defined honeycomb areas in the reflecting zones, which *appear* to be due to vesicular elevations. Within certain focal zones the centres of these dark areas reflect light.

In the case of an 18 year old girl, with parenchymatous keratitis, probably tubercular, quiescent for the past 3 years, I find the endothelium indistinct in outline and blister-like elevations in an upper relatively unaltered corneal zone. These areas may represent endothelial ablation.

With *less distinct focussing* of the endothelium and *reduced magnification* (10—24 times), we see a graining similar to that found under like circumstances in the epithelial surface of the lens. There are lozenge or diamond shaped fields, at times elevated, again giving the impression of depressions. Their diameter is about 100 microns. These fields must not be identified as cells, as has been done by others in the case of similar areas in the graining of the lens surface. Cells are much smaller.

With these brief remarks on the healthy and pathologic living endothelium, drawing especial attention to its friability and great susceptibility to changes, I hope to have encouraged further study. I will add that I have been able to easily observe the corneal endothelium in rabbits. It was in the eyes of rabbits that I first discovered it. The possibility of observing the living corneal endothelium will be of importance to physiologic and pathologic research work.

Regarding the technique, an oblique angle of observation favours the separation of the reflexes of the anterior from those of the posterior corneal surface (see Fig. 1c).

Fig. 1c. Technique of observation of the posterior corneal zone. $E =$ vertical height of entering light (radius), J, $J' =$ entering light, A, A^1, A^2, $A^3 =$ light at exit. The illustration shows that with a larger angle of incidence we can so displace the anterior reflecting zone, that it will not interfere with the examination of the posterior reflecting zone. The same radius is taken for both surfaces.

The illustration shows that with a reduced angle of incidence the anterior and posterior corneal images are too close to one another. The increasing difference of the curvature of the anterior and posterior corneal surfaces to one another toward the limbus, allows of a better separation of the images in this area (*Gullstrand*[128]). (A narrow slit will be of aid in making this observation.) The beginner had best examine (temporal light) the endothelium of a temporal zone. Let patient, with head erect, look slightly toward his nose. The *anterior* corneal reflecting image must remain visible. The axis of the microscope is approximately in the direction A, A^1. The light will fall directly on to the temporal corneal zone, unless we have approached the angle of total deflection.

We can then see macroscopically the slightly yellow reduced reflection of the posterior corneal image, nasalward of the anterior image. Focus the microscope onto the posterior image, then onto its corresponding reflecting image. *The latter is situated inside of the surface of exit (plane of exit), efgh*, Fig. 3 (compare also Fig. 1a).

With a 10—24 times magnification it shows a course granular irregular surface. With a magnification of 37—86 times we can see in it the delicate and regular mosaic of the endothelium (Fig. 3).

2*

The outline is distinct in proportion to the transparency of the corneal media anterior to it. It is advisable for the beginner to select *youthful* subjects for his investigations. The peripheral corneal parenchyma becomes slightly opaque as early as in the 30th year; this change reaching its zenith in the senile degeneration of the cornea, the arcus senilis. (Gerontoxon.)

With a little practice and by utilizing a fixed point one can easily focus the endothelial zone. The radius of the corneal area under inspection should bisect the angle between direction of illumination and observation. Various changes in the angle of observation, and the position of the patients head and eye, as well as rotation of the microscope in its horizontal axis, will allow of the examination of other areas, with somewhat increasing difficulty, because the slitlamp only permits of examination in a horizontal plane. This disadvantage is evident when attempting to locate other reflecting zones, and will call for a slight modification in the mounting of the slitlamp.

(I have been informed that the apparatus has recently been improved in this respect.)

C) REFLECTING ZONES OF THE LENS

Mention has been made that the reflecting zone of the anterior lens surface shows the anterior graining (shagreen)[20]. It also shows within these areas the fields of epithelial cells of the lens, and also the fibres of the cortical surface. The epithelium is more distinctly seen with nitrogen and arclight, than with the Nernst fibre. It is quite easy to focus the slitlamp on the anterior lens reflecting zone.

Let the patient look straight ahead, using for instance a 40° angle of illumination temporalward. The eye now should turn slightly temporalward, so that the direction of the patient's vision, or more exactly the radius of the lens surface under observation bisects the angle between axis of illumination, and that of the microscope. The axis of the latter is now in the main direction of the light reflected by an illuminated part of the anterior lens surface. It is easy to locate this area and by slightly varying the position of the light, or having the patient change his point of fixation, we may study the graining of the anterior lens surface, in its entirety.

The larger the radius of curvature of the anterior lens surface, the larger the field under observation. In the rabbit therefore the fields are much smaller.

With low magnification we see lozenge or diamond shaped fields, similar to those found on the corneal endothelium[21]).

The epithelial cells are visible only with a magnification of from 40 to 50 times, and in many cases are not as sharply defined as the endothelium. In the anterior lens reflecting zone we can easily observe the lens fibre designing of the cortical surface and the sutures (seams) (Fig. 92). These two phenomena contribute to the picture of the anterior graining.

As pathological changes in the anterior lens graining we mention the *subcapsular vacuolar layer*[21]) described by the author (droplets are found in the epithelium and directly under it, especially numerous in cases of mature cataract). Also the *Shagreen spheres*[22]) [23]), the character of which still remains to be explained.

We will now describe the iridescence in the anterior area of the lens, or rather of the anterior lens reflecting zone (Fig. 184—185). It is a phenomenon of dispersion of colour (interference) which must be accepted as a symptom of certain types, as well as of senile cataract[24]). In traumatic cataract it shows up in a special vivid and pronounced manner, if the capsule has not been ruptured (contusional cataract), or if after injury, it again has closed. If we can imagine that the iridescence is due to a thin layer of fluid (like the colours in thin flakes) we may assume that in injury to the capsule this fluid may have had an opportunity to escape, hence the absence of colour phenomena. That this phenomenon occurs only in eyeballs injured by fragments of copper, as has been recently stated, has been disproven by our observations. It is however possible that the peculiar type of cataract complicating this form of injury is characterized by a vivid iridescence of the anterior reflecting zone (*Purtscher*)[25].

This phenomenon, as well as the graining, is visible only by focussing the anterior reflecting zone.

The author[14] has also drawn attention to the *posterior lens reflecting zone*. It also shows a graining not heretofore described (Fig. 96—98).

In the middle (axial) zone, it is composed of irregular fields and tortuous lines. In the periphery the fibre formation of the posterior lens surface is distinctly seen. (In age I have observed axial as well as parallaxial, a course "plaque" or field-like pattern [Felderung].)

Deposits and other changes of the posterior lens surface show black in more luminous surroundings, as on the posterior corneal surface. According to our observations[26], a similar and often vivid phenomena of *posterior iridescence* may be seen in the region of the posterior pole and is characteristic of *Cataracta complicata*. In a less degree we find it in senile lenses and senile cataract.

I have also noted the same form of iridescence in all forms of *secondary cataract*[24] (Fig. 288, 289). Finally we must mention the reflecting zones of the lamellae (lamellar surfaces) of the lens[27] (Fig. 100a).

These show lens fibres and sutures (seams), plastic formations in age, and especially vacuole formations. These latter changes may often be seen in diffused light.

D) REFLEX LINES PRODUCED BY FOLDS IN THE REFLECTING SURFACES OF THE CORNEA AND LENS CAPSULE

Folds in Bowman's membrane and in the posterior corneal surface, as well as in the lens capsule, produce linear reflexes of a certain form and arrangement, visible with the slitlamp and corneal microscope, which we have studied with special reference to their optical character[28].

The results of these researches are as follows:

A regular wavelike surface composed of straight cylindrical reflecting surfaces, alternatively concave and convex, having similar radii, with their apices in the same plane, will, if the rays of light are parallel and vertical to the axis of the cylinder, and the observer is at a sufficient distance, produce parallel linear reflections. Their situation varies according to the angle of incidence in relation to the position of the observer.

If "ε" represents the angle formed by the angle of incidence and the perpendicular in the common tangential plane, and "β" the angle formed by the perpendicular and the direction of observation, and the sum of ε and β changes, the reflections are displaced to a similar amount.

Except for the special case that $\frac{\varepsilon + \beta}{2} = 0$ the lines appear to have formed groups of double lines. The distances of these reflections from the line which divides the space between them equal the radii of the curvatures. The distance of the lines from the top of the reflecting zones is expressed by one half of the central angle "γ" which latter is equal to $\frac{\varepsilon + \beta}{2}$.

"γ^{max}" which represents one half of the opening angle of the mirror is attained when the two lines fuse at the point of contact of the two mirrors (reflecting surfaces). If $\frac{\varepsilon + \beta}{2}$ decreases, the two lines recede in opposite directions. If $\frac{\varepsilon + \beta}{2} = 0$ the distance between all the lines is equal.

If the values are inverse, the pair of lines, now separated, approach one another, until they meet at the boundary of the two reflecting surfaces. Then the inverse maximal value of $\frac{\varepsilon + \beta}{2}$ will be attained.

If the waves become more shallow, with the same radius of curvature, if "γ^{max}", which represents one half of the opening angle of the reflecting surfaces becomes smaller, then $\frac{\varepsilon + \beta}{2}$ will decrease in proportion, and with it the area in which the lines formed by the cylindric reflecting surfaces are visible.

The convergence of the double lines at the ends of the folds is due to the reduction of the opening angle of the two reflecting surfaces. The absolute sum of the lineal displacement, due to variations in $\frac{\varepsilon + \beta}{2}$ is increased (and the lines wander the more rapidly) in proportion to the increase in the radius of curvature. With an increase in the latter we have a corresponding increase in the width of the reflection.

Dim folds appear less distinct in proportion to the degree of the diffused reflection.

These statements regarding linear reflections may be experimentally verified by artificially produced folds on smooth surfaces.

The media of the eye, which pathologically may produce these linear reflexes on their wavy limiting surfaces, are especially the cornea and retina.

The cornea in phthisis bulbi, parenchymatous keratitis (especially disciform), in perforations, operative as well as traumatic, show folds of Descemet's membrane, which latter manifest themselves by deeply situated linear opacities.

The latter, by observation with the slitlamp, show the afore mentioned reflex lines, which allow us to differentiate folds in Descemet's membrane (as also in Bowman's membrane) from similar changes, as for instance: vascular remnants, areas of clouding along nerve fibres, tears in Descemet's membrane etc.

We can therefore easily diagnose the presence of folds in Descemet's membrane with the slitlamp illumination. Descemet's membrane shows a marked tendency toward the development of these folds, especially in connection with deeply situated keratitis, and following operative perforation of the bulb.

In cases of phthisis bulbi and especially in parenchymatous keratitis the linear reflexes are irregular in form, while those following perforation are regular and radiate from the scar. According to their genesis these folds show a cross section of varied type. Folds with typic reflex lines were noted on the anterior capsule in shrinking cataracts. They may constitute the only symptom of the contraction of the lens. All of these linear reflections behave similar to ones produced artificially on reflecting wavelike cylindrical surfaces, as we have proven by experimentation on such surfaces.

4. EXAMINATION WITH INDIRECT LATERAL ILLUMINATION

If we project the image of an intensely luminous filament, for example, a glowing Nernst-fibre, onto the retina, this light is more or less diffusely reflected onto the surrounding retinal surface, causing irregularities on it to show up as reflections of varied intensity, arranged in certain directions. We may then speak of an *indirect lateral illumination*, which for example, will disclose prominences of the retinal surface, as is also attained by the methods of central ophthalmoscopy with the *Gullstrand* ophthalmoscopy, in red-free light.

(Examination with the azo-projectionlamp of *Staehli* and by the *Wolf* Method of electric ophthalmoscopy gives less satisfactory results.)

In a similar manner the bundle of light of the slitlamp allows of observation by indirect lateral illumination in the cornea, lens and vitreous.

Varying with the angle of the light as it is reflected onto the surface, we observe these tissue areas, illuminated from the front, again from the side or rear. (In the latter case we are observing by *transillumination*, which fact has been suggested by *Haab* when applied to examining the retina with the ophthalmoscope.) By the method of indirect lateral illumination with the slitlamp, we focus the microscope *onto the margins* of the luminous zone. (In a similar manner as we may make observations in focal light and by transillumination.) In this manner the vacuole formations in the areas of the anterior and posterior lens surface (subcapsular vacuolar layer) as well as on the anterior and posterior corneal surface are more distinctly seen, than by any other method.

If we examine deposits, individual cells etc., under conditions allowing *indirect lateral illumination* they will assume a plastic form, giving the impression of prominences. By neglecting to pay attention to the angle of light when examining deposits on the posterior corneal surface, one might mistake prominences for depressions of the surface. (Inversion of the relief.) Projections arising from the posterior corneal surface and into the aqueous, show as small concave mirrors, because of an increased luminosity on the side away from the source of light.

D.

EXAMINATION OF THE CORNEA AND LIMBUS*

1. THE NORMAL CORNEA AND LIMBUS

A) THE NORMAL CORNEA IN FOCAL (DIFFUSE) LIGHT
OBSERVATION OF THE REFLECTING ZONES

Fig. 1a. Schematic illustration showing the passage of the luminous bundle of light through the normal cornea (compare page 9).

$abcd$ = anterior surface of the cornea (surface of entry); bd = anterior edge, especially distinct after the instillation of fluorescin; $bfdh$ = the cut surface, especially important for the purpose of localization (shown by hatched lines). Edge eg as a rule is only visible in pathologic cases, and at times in normal corneal zones at the periphery** (senile and presenile phenomenon); $efgh$ = posterior surface of cornea (surface of exit). On this surface precipitates are found, they are not visible in the area $aecg$.

Proceeding from fh toward eg they become less and less distinct (compare Fig. 20), on account of the interposition of more dense luminous areas. For the purpose of determing the depth of a certain area, we allow the latter to enter into the cut section $bdfh$. The latter need not be visible. It is only necessary to locate bd and fh. The plane between these two edges will then represent the cut section $bdfh$. With the light from the temporal side we place the surface $bfdh$ (and with it the whole of the bundle of light) temporalward to the area in question. Now carefully approach the latter toward the surface $bfdh$ until it is *just about to enter* this plane. This localizes it in this surface and consequently also in the cornea (compare text on page 9). In a similar manner one may localize areas in the lens. To obtain a sharp picture of edge bd and fh the bundle of light and microscope must individually be in correct focus. For the purpose of determining depth we usually select Oc. 2 and Obj. a2—giving a 24 time magnification. The slit should be quite narrow.

Fig. 1b. Cylindrical bundle. (See page 10.)

Fig. 1c. The visibility of the endothelial reflecting zone. (See page 19.)

Fig. 2. Images of the normal cornea (compare page 18).

To the left the sharply focussed anterior image. To the right the yellow (olive-green) correspondingly smaller image of the posterior cornea of a normal eye.

The image of the anterior cornea shows a slight chromatic aberration at its borders (Oc. 2, Obj. a2).—(By focussing the nitrogen fibre on the diaphragm of the illuminating lens one may see an image of it in the picture.)

Note the sharp, regular borders of these pictures. Only coarse irregularities are here manifested, in contrast to the reflecting zones (Fig. 3), wherein the most minute surface variations may be easily observed.

---

* In the illustrations, the direction of the light is, if necessary. indicated by an arrow.
** By narrowing the bundle of light this edge may be made visible in normal tissue.

Fig. 3. Normal corneal tissue. A girl, 18 years of age. Normal anterior (Sp)
and posterior (Sp') reflecting zones of the cornea (compare page 18).

Both zones are *within the region of diffuse reflection (D, D')*, the anterior within
$abcd$, the posterior within $efgh$ (Fig. 1a). Oc. 4, Obj. a3.

After the reflected images (Fig. 2) have appeared in the field of vision, in order
to observe the anterior *reflecting zone*, we focus the anterior corneal surface in the
direction of the anterior reflecting image. *Sp* then appears within the stripe D, it
ends at the border ac of stripe D. Its borders are curved, not distinctly defined,
and the corners are rounded. On the superficial corneal surface movable black dots
and ringlets are seen. They are corpuscular elements in the lachrymal fluid. They
give origin to colours of dispersion (interference colours), especially after rubbing the
eyelids, and after the instillation of ointments, which increase secretion (Fig. 3). In
addition one may observe small white mosaic-like areas not of the epithelium,
movable with the surface fluid (compare page 18).

Let us now focus somewhat more deeply onto the *posterior reflecting zone (Sp'.)*
We may now see the living endothelium, as we have recently observed it for
the first time. The mosaic is composed of small, usually hexagonal fields (compare
page 18)[15)][15a)]. The colour of this zone is olive-yellow. Its borders are less distinct
compared to those of the anterior area. Especially in the periphery one may note a
flat grooved, wavy irregularity. This condition is found in all corneas and is increased in age and in pathologic eyes. The letters a to h refer to the borders of
the diffusely illuminated corneal area (compare Fig. 1a). This area shows normal
corneal tissue in focal light. We may easily recognize a delicate wart-like designing,
in which (in the chosen angle of illumination) are situated long, horizontal, indistinct,
lighter spots of varied sizes, on a uniformly dark bluish-gray background. There is
therefore no area in the normal cornea which is "optically empty". All tissue
elements reflect the light in varied degrees. This may easily be comprehended, if
we consider that the corneal substance is saturated by a fluid, which has a different
index of refraction (if we alter this index of refraction by injecting normal salt
solution, water or even air into the corneal tissues, we may instantly cause the latter
to become opaque). On the other hand, the solid elements which constitute the
corneal tissue, the epithelium, corneal lamellae and corpuscles are of a different
physical nature and therefore present varied indices of refraction. The lighter spots
are due to the fixed corneal corpuscles, while a layer formation, which may be due
to the lamellae, is not discernable. This latter however we have observed when
demonstrating the *Bowman's Canaliculi.*

Within the corneal substance, always medially and superficially, never in the
deep areas, we may observe the *corneal nerve fibres* (compare Fig. 7). Apparently in
connection we these we find circumscribed grayish-white, usually round opacities,
measuring 0,03 to 0,07 mm in diameter (compare under keratoconus, in which cases
these opacities are seen with great frequency). With a little practice it will not be
necessary to especially focus for the reflecting *images.*

Fig. 4. Amorphous endothelium, Mrs. E. age 49.

Presenile grayish hair. (Complains of asthenopia.) Vision normal, emmetropia,
no distinct arcus senilis, tension normal. Note the disseminated light spots which
give the impression of rough knob-like prominences.

Fig. 5. Senile irregularities and depressions of the cornea, Mr. Sch. age 55, peripheral corneal area.

Oc. 4, Obj. a3. Note the defects which simulate pit-like round depressions in the epithelium. They are typic of age. The pits in this case measured 0,07 mm to 0,08 mm in diameter. The endothelial design is quite distinct.

(We have proven these to be the "warts" described by *Hassal* and *Henle*[15a]).

Fig. 6. Pitlike depressions as in the preceeding case, Mrs. S. age 68.

(Hassal-Henle "Warts".) Oc. 2, Obj. a3. We may speak of these as "pits", when considering the reflecting surface a plane one, as they present a cupping on its posterior surface. These "pits" are on the posterior surface of the wart-like thickened membrane. The endothelial covering over them may at times be thin, incomplete, or absent. (Compare *Henle*[19]), *Salzmann*[16]) and others.) The endothelium is amorphous. The "pits" are often confluent.

Fig. 7. Corneal nerve fibres, Mrs. W. age 40. (Stargardt[115]*), later B. Fleischer*[55]*).)*

Oc. 2, Obj. a2. Area of a cornea, especially rich in normal nerve fibre formation.

More than a year ago this woman suffered an attack of recurring herpes corneae febrilis. We have often noted that in keratitis the corneal nerve fibres appear with greater distinctiveness (recently confirmed by Verderame). As a rule the corneal nerve fibres show a dichotomous (rarely a trichotomous) branching. The nerve trunks are thicker near the limbus. There are no corneal nerve fibres in the deep parenchyma, as can be seen by examining surface $bfdh$ of Fig. 1a, and in Fig. 7.

Often the nerve fibres are in a distinct layer over a large area, as can be seen in Figs. 7 and 1a. The two "optical sections" between which the nerve fibres are visible are the nearly parallel surfaces $aecg$ and $bfdh$ of Fig. 1a. A nerve fibre, if parallel with the corneal surface must pass through the surfaces $aecg$ and $bfdh$ in such a manner that the visible nerve ends are at an equal distance from the two anterior edges ac and bd. In Fig. 7 "t" shows a deep, and "o" a superficial nerve fibre. At K one sees two nerve fibres crossing, a single deep and a forked superficial fibre, which show a parallax. The use of the cylindrical bundle of light or a narrow slit, is of said in the determination of depth. Above there are two nerve fibres with knob-like thickenings[29]), one is situated at a point where the fibre branches. These thickenings are not at all rare.

Fig. 8. Medullated corneal nerve fibres in a 25 year old girl. (Stargardt[115]*), and others.)*

(Oc. 2, Obj. a3.) All nerve trunks show medullated sheaths in the peripheral cornea. One of these is illustrated in Fig. 7 at a, extending 0,5 mm beyond the fibre network at the limbus, with a gradually reducing medullation. (This tapering off of the medullary covering may be found quite regularly, though not seen with uniform distinctiveness.) Similar sheaths cover all other nerve trunks near the limbus.

Fig. 9a and b. Normal (juvenile) corneal and conjunctival limbus. Superficial lymph plexus (?), vascular loops and physiologic dew-like changes at the limbus.

Fig. 9a. Age 17. Lower outer corneal border. (In the following text we also refer to Figs. 10, 11 and 32.) (Oc. 2, Obj. a3.) To the right at A focal illumination. To the

Fig. 1—11.

Tafel 1.

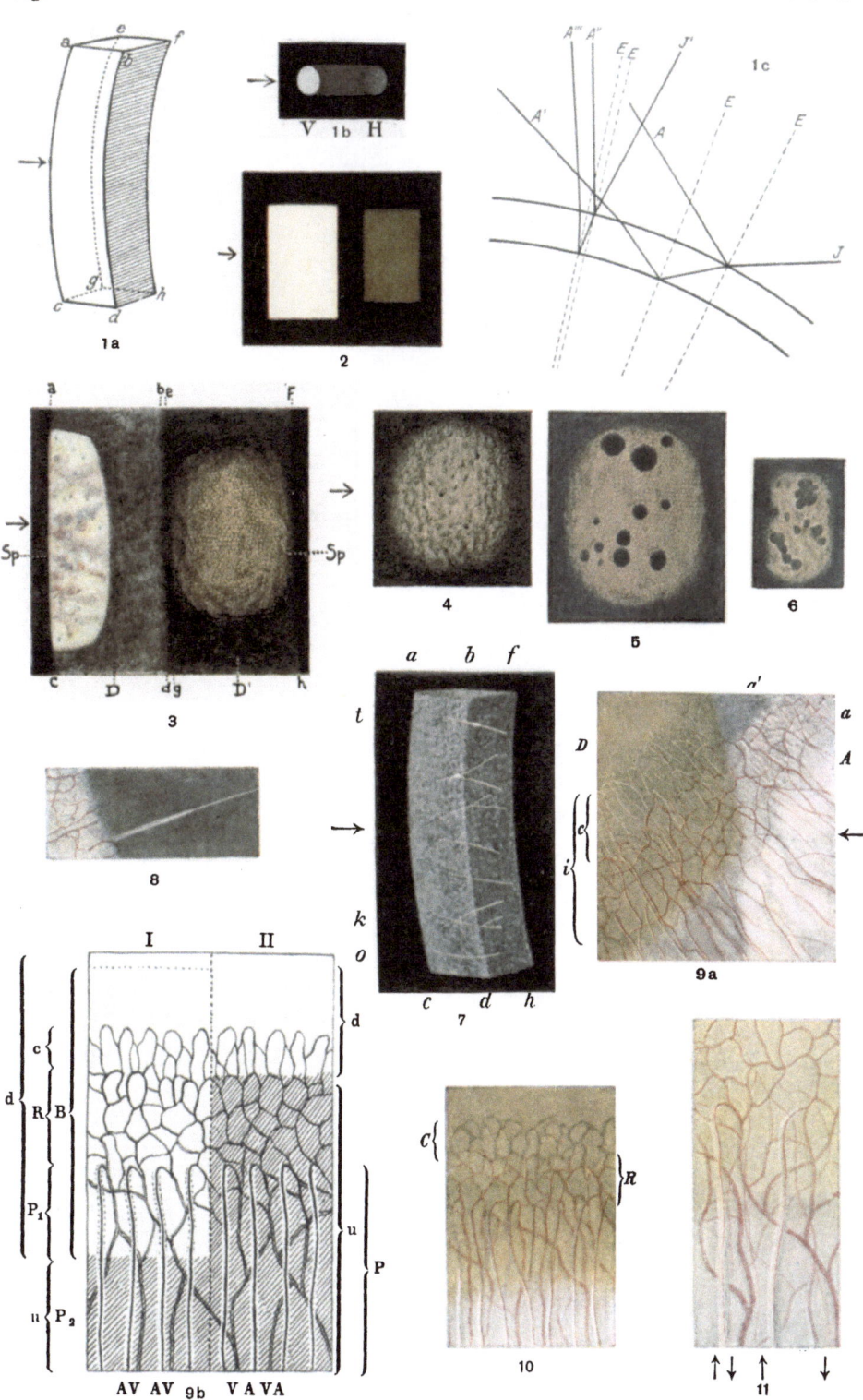

left at D transillumination. Section aa' shows the limbus in focal light, section i by transillumination. In the latter the network of vascular loops is more distinctly seen, and the circulation of the blood can be easily observed. On the other hand the (presumably) *superficial network of lymph vessels*, are better seen by focal illumination. (The white lines seen in section aa' may be considered examples of these.) These lymph-vessels cannot be seen by transillumination. The lymph sheaths of the vessels, first observed by *Koeppe*[30]) are rarely distinct. They are mostly found on veins.

The *vascular loops and network (Stargardt*[115])) present delicate angular and arched shapes, with many anastomoses (Figs. 10, 11 and 32), in which the circulation of the blood is especially distinct. I have seen individual loops in which the blood, from time to time changed in its direction *(Schleich*[116]), *Stargardt*[115])).

A remarkable formation not present in all individuals, especially observable in the upper and lower conjunctival limbus is a series of *palisades* (Figs. 10, 11 and 32), composed of radial stripes, which if pronounced may be visible macroscopically. They seem more white and distinct in age.[*]

In the case of Fig. 9a they are not visible, while in the cases of Figs. 10 and 32 they are very pronounced. These radial palisades are usually most distinct at the lower limbus and belong to the superficial conjunctival tissue.

Each palisade contains a thin blood vessel, and the direction of its blood current shows it to be a "vas. afferens". These vessels constitute the superficial arterial connection with the vascular loops. The latter in the case of Fig. 32 *project above* the palisade area in a corneal direction by 0,3 to 0,4 mm. (I have found similar conditions and measurements in other cases.) At times one may find palisades which contain no vessels. The zones of palisades do not alone vary in individuals in being more or less distinctly seen, but also present differences in width.

In the case of Fig. 32 the radial width of the zone at the limbus was 0,7 to 0,9 mm. In other cases I found them narrower. The distance between the palisades measured 0,1 to 0,15 mm. Their individual thickness in the case of Fig. 32 is 0,03 to 0,05 mm. In youth, by indirect light, they have the appearance of luminous double walled tubules. On account of their reflection we see these palisades more distinct in indirect than in direct light. In age these structures may be absolutely white and opaque. As they anastomose laterally a network is formed, into which in advancing age, pigment may be deposited (compare Figs. 33 and 34).

The relation of the individual zones of the limbus to one another as seen in direct and indirect light, in an individual 27 years of age, can be observed in Fig. 9b.

The right part of this illustration (II) shows the limbus in focal (direct) light, the left (I) by transillumination (indirect). $d =$ transparent, $u =$ the same, opaque, $A =$ arteries of the palisades, $V =$ veins, $P_1 P_2 =$ palisade zone, $B =$ zone of dewlike changes.

In the palisade zone $P_1 P_2$, which measures about 1 mm, only P_1 is translucent. At times I found the whole zone opaque as in Fig. 32. Adjoining the palisade zone is the zone of vascular loops R, which contains no palisades and appears opaque in focal light and transparent by transillumination (compare zones R in Figs. 10 and 32).

Adjacent to this area is the 0,2 to 0,3 mm wide zone of end capillary loops C, which normally are almost bloodless, quite indistinctly seen, and vary greatly in development in different individuals. (Compare C in Fig. 10, especially distinct in C Fig. 9.) Somewhat beyond their axial borders the physiologic dewlike changes disappear.

---

[*] They are probably identical with the radial stripes of J. Streiffs[135]). Apparently the radial "pseudo-cysts" of Koeppe[37]), should be here included.

While in Fig. 9a the translucent zone i of the network of vascular loops on the nasal lower limbus measures over 1 mm width (1,2 mm), its width in age averages only $^1/_2$ to $^1/_3$ of this.

This is due to the fact that with advancing age the opaque area extends axially and a part of the small end-arcades, (capillary loops) become *obliterated*. This partial obliteration of the vessels at the limbus is a regular phenomenon of senility, according to my observations, as well as in the opinion of preceding investigators.

Koeppe[138]) describes small *lymphatic loops* which extend from the end-arcades into the substance of the cornea. We could not convince ourselves of their existence, not even by the use of the micro-arc-slitlamp. We do however see absolutely bloodless capillary loops, which fill with blood as the result of a mild massage of the eyeball. The diameter of the smallest vessels of the network at the limbus measures about 10 microns. (According to *Leber's* anatomical observations, there are vessels still smaller; his measurements however were taken in eyes after death.)

The greater part of the vessels of the limbus zone of the normal eye, when at rest, contain no blood. Often we may see vessels that at times contain short columns of blood, again they are empty (*Coccius*[119]), *Schleich*[116]), *Stargardt*[115]), Donders[120])). On irritation of an area by rubbing with the eyelid, the following observations have been made: For a few seconds there is no change, then the vessels gradually fill with blood. If the massage was sufficient, the whole capillary network which was heretofore invisible appears in a distinct manner, as if it were injected with an artificial colouring matter. In older persons, in whom the network at the limbus seemed quite sparse, I was surprised at the number of vessels made visible by this experiment. Similar observations were made on newly developed corneal vessels, which in the resting eye, only in part carried blood.

The *physiologic dew-like changes* first described by the author[15a]), situated in the corneal epithelium (Fig. 9a), is notably more delicate than the same change in pathologic eyes. It is discernable in every eye, as illustrated in the area B in the left part of Fig. 9b. Regarding the technique of its observation, it is important that the light be projected into the angle of the anterior chamber. For the examination of the nasal limbus, the light must come from the tempora side, and vice-versa.

The physiologic dewlike changes are especially vividly seen in the zone of the network of vascular loops at the limbus, that is that area of the conjunctiva which extends over the cornea. These changes however extends somewhat farther into the adjoining cornea (Fig. 9a). They are composed of very minute droplets in size about equal to epithelial cells. With the small arc-slitlamp, under a magnification of 108 times the contour of each individual cell may be observed.

As the dewlike-changes extend into the cornea, its development cannot be ascribed to any individual peculiarity of the epithelium of the limbus compared to that of the cornea. An increased saturation of the peripheral epithelial layers with nutrient fluid may perhaps be a reason for its occurrence. Irregularity of the surface, due to the vascular loops, may also be a factor. I have found these dewlike changes over superficial corneal scars, which latter have existed for various lengths of time. After an instillation of cocain, or homatropine with cocain, we have seen the dewlike changes distinctly increase, and extend over the whole of the cornea.

Fig. 10. Arteries of the palisades and vascular loops of the limbus by transillumination.

S. G. age 26, right eye, lower limbus. The palisade zone in its axial half is translucent to transparent. This is the average normal condition.

Fig. 11. Two individual palisades under a 108 times magnification, illuminated by the small arc-slitlamp.

Indirect lateral illumination. The cock's-comb shaped vascular sheath becomes thinner on the axial side, as is shown in the illustration. The artery then bends dorsalward, at first at a sharp angle and disappears into the deep tissues, being lost in the capillaries of the network of limbus vessels.

Fig. 12. Lymphatic sheaths around the veins and arteries of the conjunctival limbus, Mrs. A. K. age 51.

Oc. 2, Obj. a3. The sheaths are seen with greater distinction on the veins.

They can be seen by direct lateral illumination, as well as by transillumination. The patient has had a chronic iridocyclitis of unknown origin for over ten years, with cataracta complicata. Projection good.

2. THE PATHOLOGIC CORNEA AND LIMBUS

Fig. 13a and b. Gerontoxon (arcus senilis corneae).

Between the twentieth and thirtieth years of life, at times earlier, we may observe an increase of the inner reflection of the cornea, especially near the limbus.

The reason for this may be that, though the composition of the tissue fluids permeating this media be unchanged, the refractive index of the fixed tissues (or particles thereof) may undergo alteration.

Fig. 13a shows an area of gerontoxon "G" in the case of an individual 68 years old. Direct illumination (Oc. 2, Obj. a3).

The opaque area, which we call gerontoxon as is generally known is, as a rule, distinctly separated from the limbus by a superficial *clear* layer of cornea, 0,2 to 0,3 mm in width. The *axial* border of the gerontoxon is less sharply defined. The clear area is merely on the surface, by superficial examination it may be taken for a depression. The zone "K" (Fig. 13b) reflects little, while the zone tt' on account of increased opacification reflects to a greater degree.

In addition, in the case of Fig. 13, there is a senile deposit of pigment in the network of the sclerosed vessels of the limbus (compare also the text to Figs. 33 and 34).

The senile changes in the cornea are first of all an increase in the internal reflection, in the area of Descemet's membrane at the periphery. I have found the corneal transparency at this area reduced at times before the thirtieth year of life, and the diffuse reflection increased under focal illumination. Therefore edge eg (Fig. 1a) may often be observed in the periphery, while it is invisible, except possibly with a very narrow bundle of light, in the central corneal areas.

This increased reflection of the deeper corneal layers is continuous *without interruption* to the angle of the anterior chamber. The lucid interval, as before mentioned, pertains only to the *anterior* corneal layers. The clear interval is probably due to increased nourishment of the anterior peripheral layers, through the end-arteries of the vessel network of the limbus.

The slitlamp shows the end loops to extend into the clear, but not into the opaque areas.

*Fig. 14. Anterior reflecting zone in oedema of the epithelium (compare page 12).
Case of absolute glaucoma Mr. M. R. age 50.*

Oc. 2, Obj. a2 by focal light. Similar appearances may be observed in parenchymatous keratitis, iridocyclitis and other conditions in which a stippling of the
epithelium occurs. Note the round granular surface and the curved contour of the zone.

The corpuscular elements of the lachrymal fluid and the phenomena of interference are not shown in this illustration (we see these in Fig. 3).

If the eye is closed for an instant, the irregularity of the surface disappears for
a time. They are submerged under a thin layer of lachrymal fluid. After a few
seconds they are again visible. By indirect light, that is by transillumination, these
cases show *dewlike* changes of the epithelium[15a]) (Fig. 15 and 16). The lachrymal
fluid, on closing of the lids, will *not* cause the latter to disappear.

Fig. 15. Dewlike changes of the epithelium (compare page 12).

Absolute glaucoma (the same case as in Fig. 14, which shows the anterior corneal reflecting zone). Oc. 2, Obj. a2 by transillumination.

The vacuole-like structures, visible in the light reflected by the iris (transilluminated), are unequal in size. The difference between this clinical appearance and
that of dewlike changes of the *endothelium* are shown by comparison with Figs. 19
and 20. Dewlike changes of the epithelium are quite frequently seen in the various
forms of keratitis, iridocyclitis etc. The droplets vary in size and density.

The differential diagnosis between dewlike changes of the epithelium and those
of the endothelium, in the absence of corneal opacities, can be made by focussing
alternately the corpuscular elements on the corneal surface, and the deeper precipitates on the posterior surface of the cornea, under a linear magnification of 68 times.

*Fig. 16. Dewlike changes of the epithelium in absolute glaucoma, complicating
tumour formation. Case R. age 45.*

Oc. 4, Obj. a3 by transillumination. Depigmentation of the iris in the white
area, the brown area shows an *ectropium uveae*. Large veins are present. The dewlike changes in this case are best observed over the white background. See the
chapter pertaining to the iris (Fig. 308—310), which describes the appearance of the
latter in this case.

*Fig. 17. Various types of vacuole formation of the epithelium in a degenerated
cornea.*

Mrs. Sch. age 74, Oc. 2, Obj. a2 by transillumination. Absolute glaucoma. The
superficial parenchyma is quite vascular in certain areas. Some of the vacuoles are
elongated, in this case they are often grouped in a certain relation to the vessels.
The area surrounded by the in part angular curved vascular loops is usually free
of vacuole formation. The latter are found most numerous in areas where the nourishment is poor. In *direct* light the corneal vacuoles are dark with luminous borders.

*Fig. 18. Deep corneal vessels following parenchymatous keratitis. In focal light
to the left, by transillumination to the right.*

Oc. 2, Obj. a2. The illustration shows a normal corneal area of Miss F. Th.
age 12. Two and one half years ago she suffered an attack of parenchymatous
keratitis, due to hereditary lues.

To the left is seen the bundle of light (compare Fig. 1a). Fluorescein was instilled into the conjunctival sack just before the observation, so that the lachrymal fluid on the corneal surface shows a green colouration. This allows the anterior edge bd to appear distinctly. The often less distinct posterior edge eg can now be quite easily observed. This is due to the fact that according to our observations the permanent opacities which follow this type of keratitis are situated in the deep parenchyma. The *vessels* are seen quite near the edge fh in the optical section $bfdh$, they are therefore situated in the deep parenchyma, just anterior to Descemet's membrane. We have determined that this is their situation in hereditary specific parenchymatous keratitis, with great frequency. Note the straight almost parallel direction of the vessels, which are all situated in the same layer. (Compare also *Augstein, Erggelet, Koeppe,* and others.)

The horizontal vessels are crossed at almost a right angle by vertical vessels situated just anterior to them. It seems plausible that this peculiar arrangement of the vessels is in some way related to the structure of the parenchyma.

The Bowman's canaliculi are arranged in a very similar manner, they are straight and parallel to one another in one layer, while they are crossed by other canaliculi situated in a different layer. Just as air chooses the direction of least resistance in the development of these canaliculi, so do the vessels seem to select a similar course. (Compare *Augstein*[118]) and others.)

It may be rarely seen that the vessels in focal light (to the left in the illustration) are more or less obscured by the diffusion of light, while by transillumination (to the right), they are vividly red, easily showing the circulation of the blood under a magnification as low as 24 times.

The *dewlike changes of the endothelium* (see the following figures) are especially distinct by increased magnification. For the purpose of a better view of the other detail, it is not shown in Fig. 18.

Fig. 19. Dewlike changes of the corneal endothelium in the case of Fig. 18.

Under high magnification, Oc. 2, Obj. a3 by transillumination. There are no deposits. The droplets are all of uniform size and most distinctly seen at the borders of the illuminated area, that is, partly in the shadow (compare indirect lateral illumination, page 23).

To localize the dewlike changes, for the purpose of differential diagnosis between it and similar changes of the epithelium, it is necessary (under high magnification) to focus on the corneal vessels. (By focussing the corpuscular elements on the surface of the cornea, dewlike changes cannot be seen.) The dewlike changes in this case show that the endothelium is not quite normal. By focussing the reflecting area it can be seen that its borders are somewhat blurred, and one may note pits in large numbers (compare Fig. 28). In cases of dewlike changes of the endothelium the transparency of the posterior corneal surface is greatly reduced. These areas are gray in focal light. In cases of *acute* iritis, I have at times found the dewlike changes of the endothelium arranged in *spots,* so that the *whole* posterior corneal surface appeared as if covered by snow-flakes, in focal light. By transillumination these grayish floccules appear as a carpet of delicate droplets of uniform size. In the reflecting zone the endothelium is amorphous in character. Deposits were not definitely discernible.

Fig. 20. Non-pigmented precipitates of the posterior corneal surface.

By transillumination (to the left in the illustration), and by direct illumination (to the right).

Miss S. L. age 25. Chronic iridocyclitis for the past three months. (There are individual small tubercles on the iris which are shown in the chapter about the iris.) Nasal corneal area.

To the right may be seen the diffuse stripe, due to transillumination, with its *directly* illuminated white posterior corneal deposits. These are most distinct in the area of *f h* and become less distinct toward *e g*, as they are there veiled by the diffusion of light (compare text to Fig. 1). Edge *e g* is invisible, which is a symptom of a reduced transparency of the posterior corneal wall (compare text to Fig. 18).

To the left may be seen the dewlike changes of the endothelium by transillumination, brownish in colour in the light reflected from the surface of the iris. The larger precipitates are translucent, often concentrically striped, and the edge toward the source of light is luminous.

The droplets which correspond to cells in rows and in clumps deposited on the endothelium are round or oblong, as well as club and dumb-bell shaped, or polymorphous. The very delicate carpet-like layer of endothelial droplets is not shown (see Fig. 19). The endothelial borders are irregular and as a rule indistinctly seen by focussing the reflecting zone. Note the coarse granulations which seem to present a thickening. As a rule the endothelial borders become less distinct with age.

Fig. 21. Pigmented precipitates in subacute iridocyclitis.

The latter in all probability has a tubercular basis. Z. V. age 37. Duration of iridocyclitis six weeks, Wasserman negative. Oc. 4, Obj. a3. (Right in direct focal light, the left by transillumination.) The pigment in direct light is brownish-red. The endothelial dewlike changes (to the left) are most distinct anterior to the iris. (In the yellowish-red area.)

Fig. 22. Highly pigmented areas in the state of retrogression F. B. age 28.

(Suffered an attack of bilateral subacute iridocyclitis during the past six months.) Instead of the round medium sized precipitates of two weeks ago, we today see small irregularly bordered, zig-zag edged, dark brown heaps, composed mostly of pigment. Surrounding these, visible in direct light, there are grayish-white veil-like halos, which form a network, covering all of the posterior corneal wall. The precipitates, which may be likened to the knots, are situated in this network.

By examining the grayish-white halos in indirect light, they appear as a *single* layer of delicate uniform droplets (to the right in the illustration).

THE CHARACTERISTICS OF PRECIPITATION IN TRANSCIENT SYMPATHETIC OPHTHALMIA

I have observed very delicate temporary forms of precipitation as a sign of transcient rapidly healing sympathetic ophthalmia, in two cases, which latter are further described in the chapter pertaining to the vitreous. Unfortunately at the time of observation we only made sketches of the condition, instead of exact illustrations. (Compare in contrast to this the changes as found in the vitreous, under this heading, also Fig. 346.)

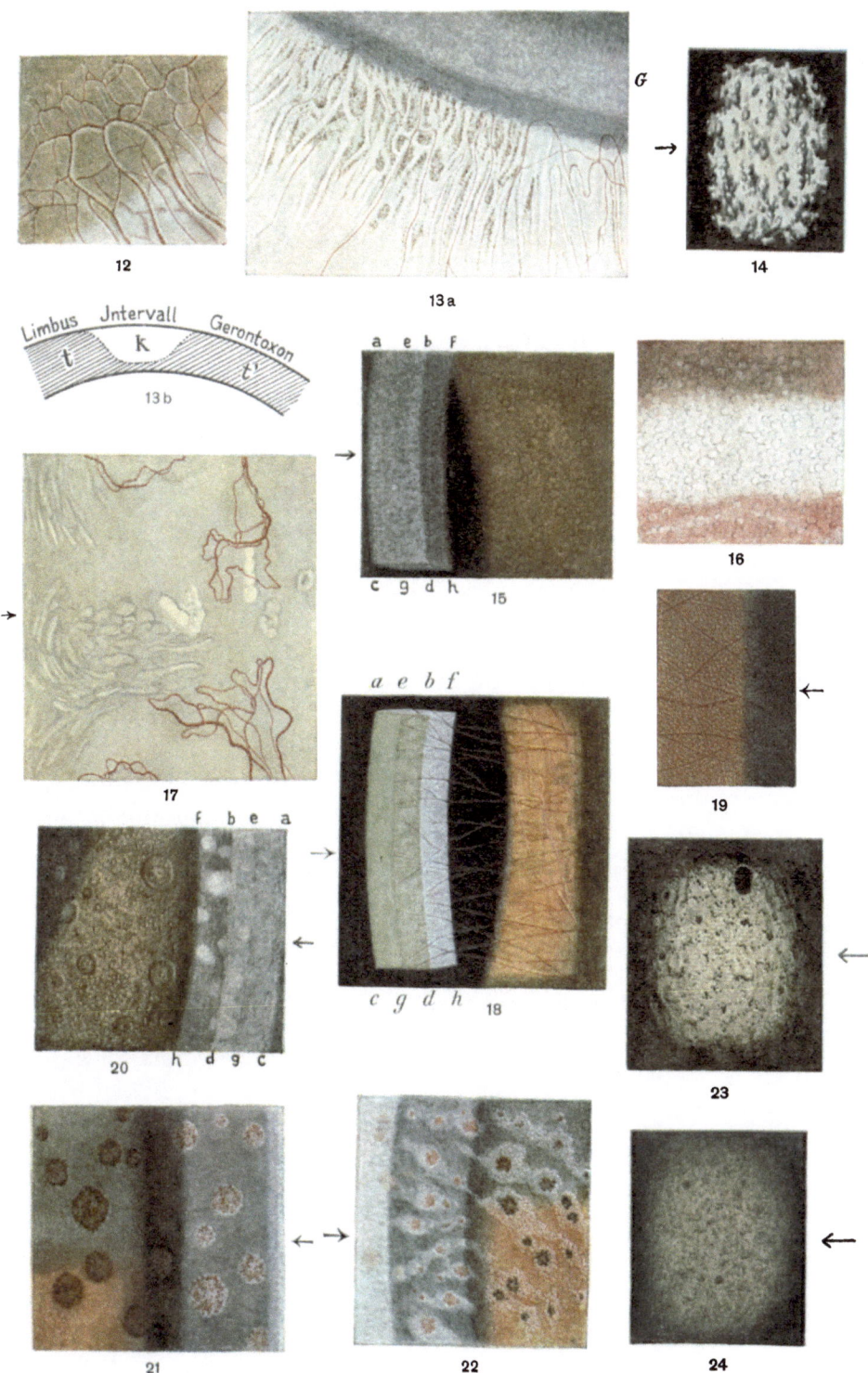

12

13 a

G

14

Limbus   Jntervall   Gerontoxon

t   k   t'

13 b

a e b F

17

c g d h   15

16

a e b f

19

f b e a

c g d h   18

h d g c

20

23

21   22   24

Both cases are herewith superficially described:

Case I. J. L. age 10, residing at S. L. sustained a perforation of the left eye by a spicule of iron. December 1ˢᵗ 1918.

Extraction by magnet after several unsuccessful attempts on the 9ᵗʰ of December. Iridectomy downward. Healed with but little irritation. Lineal extraction of the cataract. The eye was free of irritation at the time of his discharge on January 11ᵗʰ 1919. The right eye had remained normal.

On January 24ᵗʰ he reappeared, as he had repeatedly done for inspection, without suspecting trouble. It was found that the accomodative power of the right eye was reduced from a range of 12 to one of 6 diopters. With the Hartnack loupe* numerous delicate precipitates in the right (uninjured) eye. In the left cornea there are small brownish angular irregular deposits (not more than 0,02 mm in diameter), dewlike changes and deposits of individual cells. The left eye was very slightly irritated.

RS ⁶/₄ w. + 1,00, no photophobia.
LS ⁶/₁₈ w. + 11,00, aphakia.

No subjective complaints. No iritis, no synechia, no preretinal reflex lines, the macular reflex is normal (angular). In the right eye the slitlamp showed precipitates up to 0,04 mm in size in the middle and lower corneal areas. Intensive endothelial dewlike changes. Near the precipitates there were individual cells deposited in large numbers. The precipitates are grayish-white and non-pigmented. The posterior lens surface shows no deposits. In both eyes *the vitreous shows red to pink dots* in great numbers. (Compare the illustration in that chapter, Fig. 346.)

Treatment: Rest in bed, inunctions, warm applications, atropine and neo-salvarsan (according to the method of *A. Siegrist*) intravenously 0,15, later on 0,2 and 0,3.

The following days showed a retrogression in the corneal precipitation. On Feb. 7ᵗʰ 1919 only exceedingly small, hardly measurable pigment granules were found present. A week previously they were white and free of pigment. On the other hand the precipitates in the vitreous have increased, especially in the lower area. They were absent in the retro-lental area. (In a small quadrilateral area measuring less than ¹/₂ mm on Feb. 11ᵗʰ 1919. I counted 15 to 20 small and 3 or 4 large plates, the latter apparently up to 0,02 mm in size.)

(With the red-free light the vitreous showed delicate dust.) The whole supporting structure of the vitreous when observed with the slitlamp was increased in luminosity, compared to its former appearance. The luminous dots lie on the supporting structure. Even when they seem to be free of attachment in a constant unchanging position, we must presume that they are supported by invisible fibres or lamellae. (It cannot be said that the vitreous is "optically empty"•if we fail to observe a supporting structure by a certain method of examination. In areas where I failed to see these structures with the Nernstlamp I have often found them present by examining with the micro-arc-slitlamp).

In the following week the balance of the small corneal precipitation had disappeared completely. In the left (injured) eye the endothelial dew-like changes and pigment dust of the posterior corneal surface was still distinctly visible. Vis ⁶/₆ w. + 11,00.

The patient left the hospital on March 8ᵗʰ 1919. Endothelial dewlike changes. Some vitreous dots in the right eye were still found present after four months. Today on August 9ᵗʰ 1919, by careful examination, not the slightest sign of them was visible. The left (injured) eye however still shows endothelial dew-like changes below, and dustlike pigment on the posterior corneal surface. Vision as before. A year later there was no change.

Case II. Master carpenter H. V. age 38, on Nov. 7ᵗʰ 1918 sustained a large perforating wound of the left nasal cornea, by being struck with a large piece of wood. Nasal corneal flap wound.

Haemophthalmos, iris prolapse R. V. = ⁶/₅ L. V. = projection of light. Healed under constant irritation Nov. 30. Removal of the iris prolapse and linear extraction.

On account of the constantly recurring photophobia, pain and irritation of the injured eye, I repeatedly suggested its enucleation, which latter was refused by the patient.

---

* Translators note: I am quite sure that the *Hartnack* spherical loupe corresponds somewhat in nature to what is known as the "Coddington".

Discharged unhealed on Nov. 28th 1918, ciliary injection in left eye, dewlike deposits, individual precipitates and folds in Descemet's membrane, vascularized corneal flap, with partially incarcerated iris.

Right (uninjured) eye accomodation 6—7 D. very slight irritability and photophobia, epiphora following exposure to light, at times a conjunctival injection. On the 5th, 9th, 12th and 28th of Dec. 1918 and on Jan. 6th 1919, the conditions were similar with less irritation. On Feby 5th 1919, after he had returned to work, the conditions were apparently similar. The left (injured) eye still showed fine precipitation and endothelial dewlike changes, no irritation.

With the corneal microscope and slitlamp the *right cornea* which had previously repeatedly been examined at intervals now shows in its lower third, a yellowish white precipitate measuring 0,04 mm. Surrounding this (visible by indirect light) "delicate fibrin-fibres" and droplets. The (right) *vitreous supporting structure* is covered by a number of white to reddish dots, as seen in the preceding case, and as are found present in cyclitis and choroiditis. Enucleation of the injured eye and hospital treatment were refused. Two days later the vitreous dots were distinctly seen and were more numerous.

By focal illumination with the ophthalmoscope in red-free light there were delicate dust-like deposits visible in the vitreous. No irritation, accomodative range unchanged. Nasalward of the right macula there were 3 or 4 preretinal reflex lines. Right macular reflex absent. (Left the same. Projection good.)

On account of an attack of influenza-pneumonia the patient was not again seen until March 12th 1919. The right cornea was perfectly clear—accomodation 6—7 D. The vitreous condition was the same. Since then the eye has been free of irritation.

The vitreous still presented deposits on its supporting structure after a further period of three months.

The injured eye was absolutely free of irritation, however on June 19th 1919 it still showed endothelial dewlike deposits. Aug. 25th 1920, no further irritation.

These two cases show:

1. *Mild ephemeral attacks of sympathetic ophthalmia may occur*, which are so free of symptoms that they were formerly rarely diagnosed.

2. It is necessary after perforating injuries to carefully control and examine the cornea and vitreous of the uninjured eye with the slitlamp and corneal microscope. When examining the cornea by *transillumination*, the pupil should *not be dilated*, so that the iris may present as large a reflecting surface as possible. Especially important and necessary is a careful survey of the vitreous, particularly *its lower areas*.

3. It is still undecided whether or not socalled sympathetic irritation is at times a genuine factor in sympathetic ophthalmia.

Fig. 23. Posterior corneal reflecting zone in chronic iridocyclitis.

Case of Fig. 20, however the posterior corneal reflecting zone. Observation of the temporal corneal area. Oc. 2, Obj. a3. Endothelium somewhat indistinct, probably due to oedema. (Compare the dewlike deposits in this case, shown in Fig. 20.) The deposits, individual cells, chains of cells, small and large clumps are *black*, and distinctly outlined. The individual cells are often arranged in rows and groups. (They are indistinct or may be invisible in diffuse light.)

Around the precipitates there may be seen a luminous border of endothelial cells, probably a change in the surface curvature. Dense extensive precipitates reflect sufficient light so as to show white in the reflecting zone that is more luminous instead of darker, compared to their surroundings.

Fig. 24. Case of Fig. 22. Posterior reflecting zone in chronic cyclitis. Miss F. B. age 28.

Oc. 4, Obj. a3. Endothelial dewlike changes and deposits. Slight increase in tension. The endothelial cell borders have practically disappeared.

Three months later, during the stage of resorption of the precipitates, there was a slight increase in the visibility of the endothelial cell borders. Extensive, distinctly visible iridescence of the endothelium, very similar to that shown in Fig. 70 in the area of a rupture of Descemet's membrane.

Fig. 25. Precipitates and surrounding cells by indirect lateral illumination (on the border of a direct or indirectly illuminated area). Miss G. age 29.

Chronic tubercular iridocyclitis. Oc. 4, Obj. a3. The luminous small globules are probably oedematous cells.

Fig. 26. Precipitates and surrounding cells in indirect lateral illumination. Miss B. age 18.

Oc. 4, Obj. a3. Chronic iridocyclitis of an unknown, probably tubercular origin. Illumination from the left side, the shadows are also on the left side. *Anterior* to the precipitates one may also see a suggestion of individual small globules.

Fig. 27. Cicatricial irregularity of the posterior corneal surface. Observation of the reflecting zone.

Oc. 4, Obj. a3. (Patient age 26. Siderosis bulbi. Perforation of the cornea two years ago.) Diagonally horizontal perforation scar.

The irregularity is recognized by the distortion of the reflecting zone, due to changes in the curvature of the posterior corneal surface, near the scar. Endothelial cells are visible in parts of the contracted area. Distortions of this nature are regularly found in cases of deep scars of the cornea.

Fig. 28. Pitting in the endothelial reflection (posterior prominences), following parenchymatous keratitis.

(Reflecting zone in focus. Oc. 4, Obj. a3.) After parenchymatous keratitis I have often noted circumscribed round dark areas in the reflecting zone. The bases of these areas give the impression of irregular dorsalward granular pitting by variations in the angle of incidence and direction of observation. I have seen very similar changes in senile corneas. In this case they correspond to the *Hassal-Henle* "warts" (see Fig. 5 and 6). The pit-like depressions in both instances measure from 20 to 100 microns in diameter.

Fig. 28 shows these protrusions in the cornea of Master E. K. age 8, congenitally luetic, who suffered an attack of bilateral parenchymatous keratitis. Today the eyes are free of irritation. The posterior corneal lamellae show a light diffuse clouding. Numerous vessels, containing blood, situated just anterior to Descemet's membrane, are distinctly seen in sharp contrast to the endothelium (see illustration).

On the side toward the light, the vessel walls show a luminous stripe. The largest vessels measure 10 to 20 microns in diameter. Several pits in the centre and above show as *dents* directed backward when observed within a certain direction of illumination. (Compare Fig. 5 and 6.)

Fig. 29. Endothelium, amorphous in character, in the area of the conical point in kerato-conus. Mrs. S. age 35.

The reflecting zone principally shows *round* dark pits, with a luminous border. By changing the direction of light one may observe the endothelium in the bottom of the depressions. This differentiates between depressions and deposits, which latter are also dorsal elevations. We are referring to Henle's warts.

3*

Fig. 30. Peculiar annular and circinate surface changes in the endothelial zone.

The picture of this reflecting zone, at first sight, permits us to presume that we are observing a flat vesicule-like bending in the endothelial reflecting surface.

Miss E. Sch. age 30. Oc. 2, Obj. a3. Three years ago suffered an attack of parenchymatous keratitis of the left eye, on a probable tubercular basis, now quiescent. The upper corneal area, which presents an irregular endothelial surface, is still relatively transparent. The endothelial cells are quite indistinct.

According to the angle of incidence of the light the rounded areas show dark or may present a luminous sharply bordered field in their centres. At times only half of the borderline of a field is visible (compare the two middle fields). This illustration distinctly shows an *irregularity* of the posterior corneal surface, while the suggestion of vesicle formation is doubtful. I have observed similar changes of the surface in deep corneal scars.

Fig. 31. The limbus in a case of keratitis superficialis scrofulosa. Soldier age 21.

Oc. 2, Obj. a3. Two years ago the patient suffered an attack of superficial keratitis. Recurrence six weeks ago. Today the eye is practically free of irritation. Epithelium smooth. Extending into the superficial corneal substance for a distance of 1 mm and more there are vascular loops, which form a fine network.

The apices of the loops are elongated and pointed. Some are filled with blood, others only in certain areas.

Fig. 32. The lower limbus in subacute iridocyclitis. The limbus also shows a distinct zone of palisades.

Mr. F. B. age 28. Case of Fig. 22. Oc. 2, Obj. a3. During the past six months suffered a bilateral subacute attack of iridocyclitis, with dark-brown precipitation and intermittent tension, making it necessary to perform bilateral iridectomies, and two anterior sclerotomies.

The limbus network of vessels is well filled with blood and from it in certain areas there may be seen, short *broad* loops with their ends flattened, extending 0,2 to 0,3 mm into the superficial corneal substance. Individual small loops extend into the *deep* parenchyma, one follows a *nerve trunk.* (To the left in the illustration.) The loop is 0,8 mm in length. The two vessels are 0,03 mm apart. The very pronounced zone of palisades (described in the text to Fig. 9) is found in normal eyes. It may be mentioned that in the area of the sclerotomy scar, the blood vessels of the individual palisades have disappeared. In their places there is brownish pigment debris in the canaliculi (see illustration to the right, below). It is very likely of hematogenous origin.

Fig. 33. Senile changes in the limbus, with pigment deposits in the limbus and in the adjoining cornea.

Mr. B. age 79. Oc. 2, Obj. a3. Direct illumination. Right lower corneal border. The limbus is surrounded by a white meshwork in which the palisades, as described in connection with Fig. 9, are visible. In the meshes of this network there may be seen brownish yellow pigment. The latter is *anterior* to the vessels and extends into the cornea. The cornea shows a well developed gerontoxon, which is separated by a 0,15 to 0,2 mm wide clear interval from the limbus. The pigment extends into the clear interval as well as into the gerontoxon. Pigment deposits of this and of similar nature are common senile phenomena, however the pigment as a rule is in the meshwork of vessels, not anterior to them, as in this case. It is probable that the pigment in this case, in part at least, is deposited in the epithelium.

Fig. **25—37**. Tafel **3**.

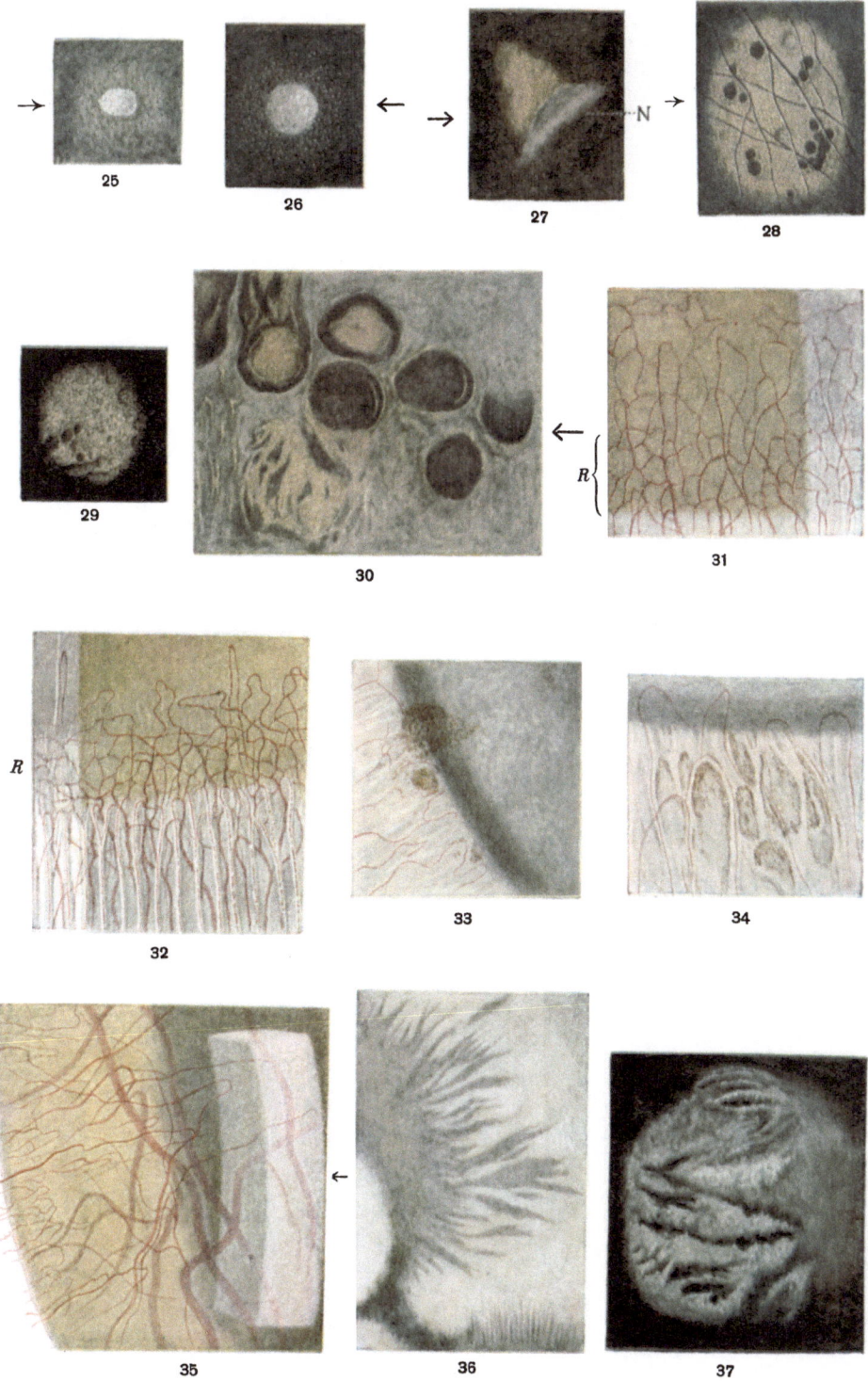

Fig. 34. Senile changes in the limbus, with pigment deposits between the meshes of the sclerotic vascular network of the limbus.

Oc. 2, Obj. a3. Senile pigment deposits of this nature, often less extensive, are frequently observed with the slitlamp. Its arrangement seems to indicate that it is of local hematogenous origin. Its appearance probably bears some relation to the senile partial obliteration of the network of vessels of the conjunctiva and limbus. (Compare also Fig. 13 which shows the limbus of the other eye of Mr. B. in good health at the age of 68 years.)

Normally one may find similar pigment granules of the basal and other *pigment cells* at all ages (for instance compare Virchow [31])-Graefe-Saemisch handbook). It however has seemed impossible to see this physiologic pigment with the slitlamp.

Fig. 35. Vascularized corneal area in a case of keratitis and chronic iridocyclitis of unknown etiology, existing for the past three years. Mrs. Sp. age 44.

Oc. 2, Obj. a2. The whole of the cornea in all of its parenchymatous layers shows a proliferation of large vessels, with many ramifications. In the area of direct illumination (to the right in the illustration) the vessels seem to be somewhat veiled by the diffusion of light.

Fig. 36. Fuchs' stripes due to clearing up of corneal opacities (Fuchs'sche Narben-aufhellungsstreifen) by slitlamp illumination.

(Observation in focal light.) Labourer L. age 25. Very old scars in the deep corneal parenchyma, which have brought about an astigmatism of a high degree.

Under a magnification of 24 times (Oc. 2, Obj. a2) one sees flamelike tongues presenting a clearing up in the scar formation. They are at times radial and again parallel in arrangement (compare below in the illustration). They are rarely arranged in an irregular manner. According to our observations, these areas showing a disappearance of scar formation indicate that the cicatrix has existed for a long time. The scars are usually in the deep areas.

Regarding the genesis of this late clearing up we can only make presumption. There is a striking similarity between the lines in this picture, and those formed by *folds* in the posterior corneal surface (for instance, compare Fig. 77).

Fig. 37. Very old central corneal opacity of the deep parenchyma. Mrs. M. B. age 66.

Oc. 2, Obj. a2. When four years old suffered an attack of bilateral keratitis (very likely scrofulous).

Note the lines in the scar area which seem to suggest a separation. Especially dense scar areas show white round spots, probably due to retrograde metamorphosis. They produce a sugar-coated appearance of the lineal opacity. At the side of this opacity I found a *superficial* macula, evidently of like age (not shown in the illustration), which shows neither the clearing-up lineal stripes, nor the sugar-coated appearance.

Fig. 38. Changes in the corneal border resembling band-shaped opacity. Mr. M. Sch. age 55.

Left eye. (The right eye presents a healed hypopyon-keratitis.) By lateral illumination one may observe a chalky white irregular peripheral change in the superficial corneal area. It is best developed in the (interpalpebral) areas left uncovered by the lids, however it is present in all of the periphery, with the exception of a few irregular areas, in a similar manner as seen in cases of band-shaped keratitis.

By reflected light one may observe, that it is separated from the sclero-corneal junction by a lucid 0,05 to 0,1 mm wide interval.

The width of the opacity is 0,15 to 0,2 mm at a nasal area. The conjunctival vessels near the cornea are in part obliterated and show two punctate hemorrhages.

The other eye presents the same changes at the limbus, which latter I have found quite frequently in senility. One must not confuse this superficial degeneration with *gerontoxon*, which latter is very different in appearance and at the same time, in addition, involves the deeper parenchyma. The change must be classified with band-shaped keratitis, in which latter condition I have at various times seen similar variations. (Deposits, in all probability similar, are mentioned by *Koeppe*[27]).

Fig. 39. Band-shaped opacity of the interpalpebral corneal zone. Appearing in the central interpalpebral aperture, six months after the beginning of sympathetic ophthalmia in Miss E. H. age 5¹/₂. (This case is described in the chapter pertaining to the iris, Fig. 317.)

Oc. 2, Obj. a2. The fairly well circumscribed elongated. oval opacity is situated in the superficial zone, probably near Bowman's membrane. It is of similar texture, though showing numerous dark round areas. Note that the surrounding cornea, especially above, is not normal. It shows a delicate superficial marble-like opacification. Nasalward there are a few isolated opaque stripes.

The eye is free of irritation. Two weeks ago the tension was reduced. (At this time, after regular weekly inspection extending over a long period, the band-shaped opacity was first noticed). The use of atropine raised the tension to normal. There is incipient cataracta complicata.

Fig. 40. Clouding of the cornea of a band-shaped variety in hereditary hydroph-thalmos in a rabbit, 12 weeks old.

(The hydrophthalmos of this animal, which also existed in the other members of the same litter, and of their progeny, will be more fully described elsewhere.)

The clinical picture also involving Bowman's membrane, resembled that of band-shaped keratitis in human corneas, but extended still less far laterally toward the limbus. In all cases observed in this particular breed of rabbits the form in which the opacity occurred was similar. Just as is seen in the same opacities in human eyes, these rabbits' corneas presented lucid stripes, which resembled ruptures in Bowman's membrane (dark in direct light). They increased in number and in their width during an observation period of one month in this case.

Fig. 41. Incipient band-shaped keratitis.

Present in elderly people, but owing to the slight degree of involvement often overlooked, is a band-shaped keratitis such as is quite commonly seen in cases of shrunken eyeballs. In its incipiency it can often be diagnosed by the slitlamp. Near the limbus, but separated from it there are changes as shown in Fig. 38.

The opacity situated in the palpebral aperture is shown in Fig. 41. (Mrs. Sp. age 84, eyes otherwise quite normal, Oc. 2, Obj. a2.) It is separated from the altered sclero-corneal junction by a *torn-like* slighty opaque zone on both sides. Below, a vessel loop extends for quite a distance beyond the limbus network of vessels into the opacity.

Fig. **38—46.** Tafel **4.**

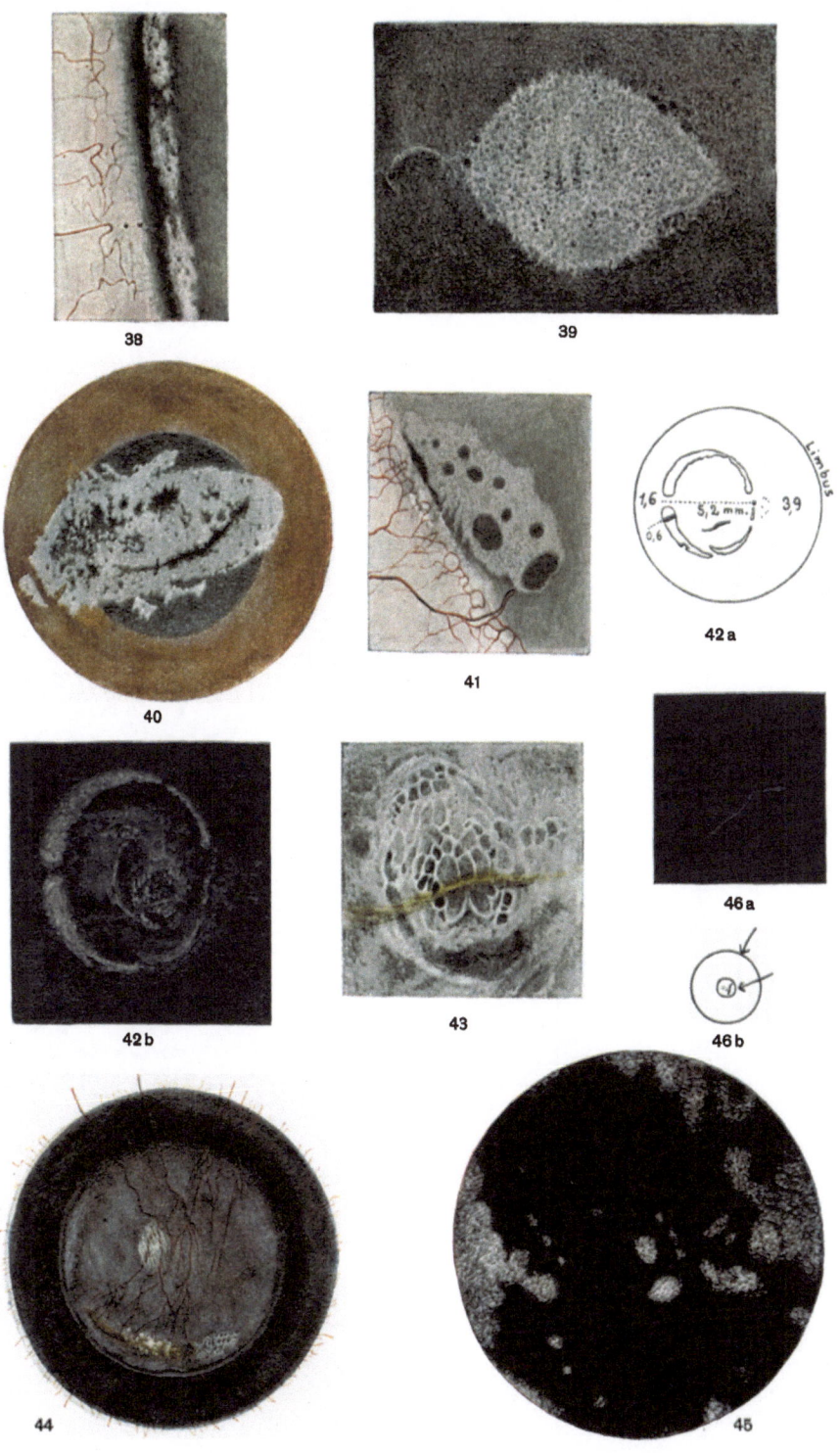

The numerous round transparent areas give a sieve-like character to the opacity. Both eyes show a senile scattering of pigment in the anterior chamber, with corresponding alterations in the iris.

Fig. 42 a and b and 43. Brownish-yellow pigment line * and honey-comb-like changes in a cicatrix following discform keratitis. Mrs. Sch. age 48.

Keratitis 5 and 4 years ago. $RV = \frac{1}{10}$. The scar is discformed, gray marble-like with a grayish white border. The measurements are given in sketch 42a (measurement under a 10 times magnification with ocular-micrometer 2).

Fig. 42b represents the cicatrix under, a 5 times magnification. Fig. 43 shows the yellow line and its honey-comb-like surroundings under a 24 times magnification in focal light.

The honey-comb areas have flattened, often hexagonal borders. The yellow superficially situated pigment line extends somewhat into the walls of the honey-comb areas. It lies in the interpalpebral space.

Honey-comb-like changes of this type should not be confounded with epithelial vacuoles such as are shown in Fig. 17.

We have found these changes quite frequently in recent as well as in old cicatrices. (Changes similar in type are mentioned by *Koeppe*[139]), who claimed them to be cysts.) The yellowish brown pigment line, usually situated in the palpebral fissure is, as is known, quite often seen in old corneal scars.

Fig. 44. Pigmentation and honey-comb-like (cystoid) changes in an old corneal cicatrix.

Under a magnification of about five times in focal light.

Mr. B. age 58, developed a discform (?) keratitis four years ago. Just as in the preceding case we find a discform, almost central opacity. The cicatrix is richly vascularized from above. In the central area there is a brilliant, rather superficial fibre designing (crystallic needles?). Below this a nearly horizontal greenish yellow superficial pigment line, and to its right a honey-comb-like area similar to the one seen in the preceding case.

Fig. 45. Corneal opacities in their form characteristic of several types of herpes zoster ophthalmicus.

Mrs. M. age 60. Herpes zoster ophthalmicus six months ago. Involvement of the cornea and iris. Today the eye is quiescent.

In the superficial parenchymatous layers one may observe round cloud-like, disseminated, often confluent flat opacities, quite uniform in size. I have never seen this type of opacification except in cases of herpes zoster ophthalmicus. The infiltrations which precede this opacification are not, as a rule, complicated by an ulcerative process.

Fig. 46 a and b. Brownish yellow pigment line in an old corneal scar.

Miss G. V. age 53. Oc. 2, Obj. a2. Focal light. Old diffuse (scrofulous) corneal macula of the palpebral zone, showing a diagonal, at its ends branched, angular curved 2 mm long, brownish yellow pigment line.

---

* Regarding the pathologic anatomical basis of degenerative changes in corneal scars, compare *Greeff*[17]).

The width of the line is 0,05 to 0,07 mm. Sketch *b* shows the location of the line and its position in relation to the corneal diameter and pupil.

Fig. 47 and 48. Pigment line of the corneal epithelium as described by Staehli[32]).

This line is quite frequently seen beyond middle life and in old age, even in otherwise normal eyes, which latter show no corneal cicatrices. It is perfectly straight or may be wavy in contour. This pigment line is situated in the area and direction of the palpebral fissure somewhat above the line of the lower lid when the direction of vision is straight ahead. At times we have observed white dense dots in this pigment line (Fig. 48—50).

It is easily seen under a very low magnification in the bluish light of an azo-projection lamp. With the yellowish light of the Nernstlamp it contrasts less distinctly. It is better seen with the nitrogen or micro-arc-slitlamp.

Staehli identified it as in nature similar to *Fleischer's keratoconusring*. According to *Staehli* this line in all probability is composed of pigment granules of a hematogenous nature, which are situated in the basal epithelium.

Fig. 47 shows the line in the left eye of labourer H. Th. age 53. The small macular spot was covered by a foreign body imbedded in the cornea seventeen years ago.

Fig. 48 represents an extensive pigment line in one eye of Mr. R. L. age 54. The smooth wavy line in places shows a flat kinking and here and there punctate white dense spots. Above the line there is a small round macular spot. According to the patients statement his eye was never injured or inflamed.

Fig. 49. Staehli's line under high magnification, observed with the micro-arc-slitlamp.

(For this substitution I am indebted to *Prof. Henker* and the firm of *Zeiss*.) The intensive bluish light of the micro-arc-lamp made the examination of this line in all of its finer details possible.

Fig. 49 presents the temporal part of the line shown in Fig. 48 in exact focus, with the micro-arc-lamp illumination as just mentioned. (Oc. 2, Obj. a3, magnification 37 times.) The line at one place is forked, the lower branch is gradually lost. The width of the line *averages 0,05 mm.* Its yellow tinge gradually disappears into the surrounding tissues. Here and there are circumscribed areas of increased density which are yellowish-white in colour. The individual dots which form the line are quite brilliant. In certain areas it seems probable that the pigment extends into Bowman's membrane and the superficial parenchyma.

Anatomical investigations will probably be necessary to definitely determine whether this presentation is correct. That this yellow line may be of a transcient nature I have had occasion to note in a case following cataract extraction.

After the usual flap extraction, the formerly very distinct line *had disappeared* (Fig. 50). In the light of the micro-arc-lamp the former site of this line could be identified as a pale bottle-green, indistinct stripe. Could it have been scraped off with the epithelium?

The arc-lamp naturally enhances the distinctness of all corneal detail. The nerve fibres are especially conspicuous and show their finest ramifications under high magnification. They are quite luminous, as if composed of brilliant dots. (In the illustration one may observe two nerves passing under the yellowish pigment line.)

The endothelium in all of its detail, as well as the supporting structure of the vitreous are seen with a surprising distinctness.

It is a question whether or not this intense illumination is injurious. I have not observed any irritating after-effects. Various rabbits, in whom *Dr. U. Lüssi*, assistant ophthalmologist at the clinic, *continuously* exposed a circumscribed corneal and lens area for three quarters of an hour, showed no changes. Evidently the invisible rays are greatly eliminated by the glass lens. In the meantime we will only use this method of illumination exceptionally, and for short periods only, until observations have definitely proven that its use is absolutely safe.

Fig. 50. Staehli's line under a 68 time lineal magnification. Illuminated with the micro-arc-slitlamp.

Housemaid Sch. K. 68 years old. Bilateral senile cataract. *Staehli's* line is similarly present in both eyes. It is situated below the corneal centre in the pupillary area. It is composed of dots which vary in the density of their grouping. Note the increased density in the pointed zones, and the characteristic kinking to the right in the illustration. The width of the line varies, as may be seen.

In certain areas of the line there are delicate opacities of the tissues, probably in the vicinity of Bowman's membrane, or in the superficial parenchyma. The intensively white areas in all probability are not due to pigment deposits in the tissues. To the right in the illustration, above the line, there are three or four small, indistinct dots of an unknown nature in the parenchyma. Near these are two nerves, one of which divides into branches.

Fig. 51 and 52. Annular traumatic corneal opacity. Form I.

This opacity is in all probability often overlooked when examining under focal illumination. It is observed after circumscribed corneal contusion, especially following injuries due to explosions.

The case illustrated by Figs. 51 and 52 is of a woman 34 years old, who was injured by the explosion of a cap. At three places small particles entered the superficial cornea. One of these is in the superficial limbus on the temporal side and the small red spray above and to the right is a conjunctival hemorrhage.

An annular opacity with a diameter of about 2,5 mm has formed around each foreign body. The *diameter* of the ring surrounding the spot at the limbus is greater (radius 2 mm), however one may observe a second smaller ring about 1,5 mm in diameter concentrically situated within the larger one. The two temporal rings are united, so that they together form an elongated biscuit. The *width of the rings* is about 0,25 mm.

These rings are found in the *middle parenchyma*, while the foreign bodies are all situated *superficially*. Within the rings the parenchyma shows *striations*, similar in appearance to those seen in keratoconus. They are probably separations of lamellar layers and are comparable to Bowman's Canaliculi. In the ring centres one may see striations of the kind mentioned *crossing* one another, situated however at *varied* depths.

Fig. 52 shows the middle ring under a magnification of 10 times. The diameter of the ring is 2,65 mm. The length of the line is 0,5 mm. The single line in this ring is situated 0,8 mm distant from the temporal upper ring border and 1,3 mm from the lower nasal border. In the centre one may see the white luminous superficially situated largest foreign body.

The ring formation had fully disappeared fourteen days after the injury.

The striations in the parenchyma persisted for weeks, some as long as four weeks, they then also disappeared. The smaller foreign bodies were retained without irritation.

Fig. 53 and 54. Annular traumatic corneal opacity. Form II.

G. K. foreman in a stone quarry, sustained an injury during an explosion on the 7th of May at two o'clock in the afternoon.

On the 8th, at ten o'clock in the morning, several minute particles of stone were found imbedded in the superficial parenchyma of both eyes. Around *every particle* were seen concentric ring formed opacities of the *deep parenchyma*, with a corresponding *annular forward bulging of Descemet's membrane*. In addition there were sparce brownish deposits on Descemet's membrane within the ring area. The diameter of the circular opacities in the right cornea were 0,2, 0,4, 0,5, 0,8 and 1,1 mm respectively (Fig. a).

A larger discform opacity (2,5 mm in diameter) was observed in the left eye (Fig. 54 above). It was in the deep parenchyma.

There were individual folds in Descemet's membrane and light and dark stripes crossing one another. The foreign body was situated excentrically upward in the superficial parenchyma.

On the following day most of the rings had completely disappeared, a few were faintly visible.

The extensive opacity in the upper part of the left cornea was still visible after a few days. This second form (Figs. 53 and 54) differs from the first form of temporary corneal cloudings (Fig. 51), especially in the fact that they were situated in the middle parenchyma in form I and in the deep in form II.

In the latter type Descemet's membrane seems to bulge forward in a manner corresponding to the shape of the ring, and the duration of the opacity, compared to that of form I, is more *temporary in nature*. The opacities of form II were visible only by slitlamp illumination. In both cases the very minute foreign bodies were imbedded *very superficially*.

In these cases it seems that the appearance of the opacities is due to the intense circumscribed shock which has produced a momentary indentation and a kinking of the cornea at the ring area. The kinked area seems to suffer most, hence the circular opacity. Why the latter in the one case was situated deeper in the parenchyma than in the other case cannot be decided at this time. (Compare the observations of *Meller*[33]), *Caspar*[34]) and *Pichler*[35]) regarding annular traumatic opacities of the cornea.)

Fig. 55. Peculiar coloured deposits on the posterior corneal wall in chronic kerato-iritis of unknown etiology.

Focal light (Oc. 2, Obj. a2). Mrs. A. H. age 56.

In childhood and again ten years ago she suffered an attack of superficial and deep bilateral keratitis, complicated by a secondary iridocyclitis. There are old, in parts vascularized opacities and a high degree of irregular astigmatism. Wasserman negative. There is no history of trauma.

Three months ago a recurrence of the keratitis occurred, accompained by chronic iridocyclitis. There were individual deep corneal infiltrations and precipitates. In the lower third of the cornea, we noted for several months, a coloured, closely meshed

Fig. **47—54.** Tafel **5.**

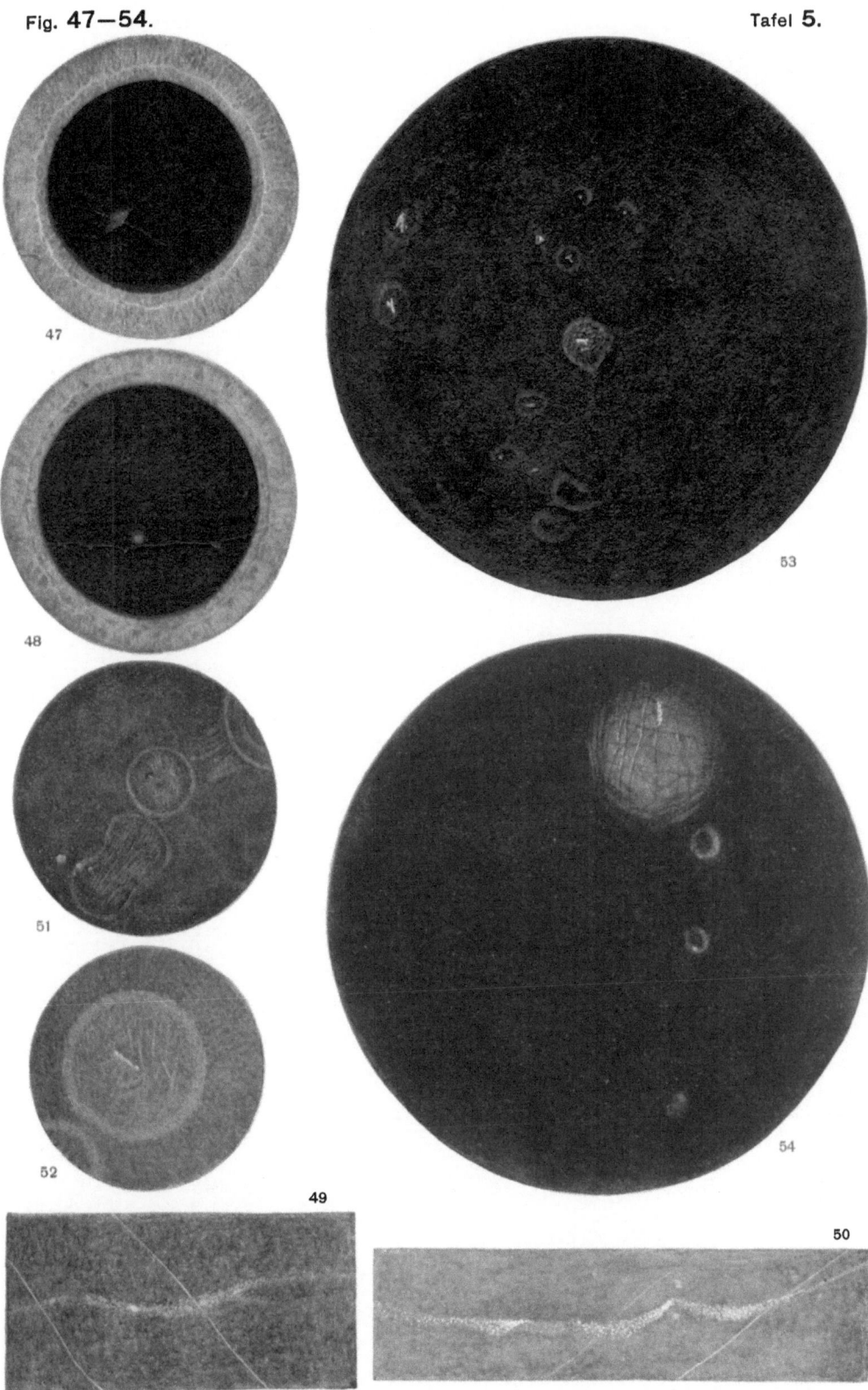

47

48

51

52

49

50

53

54

network of deposit, which gradually disappeared. Its upper part was vividly *ultramarine blue*, its lower part *bright yellow* to slightly greenish yellow in colour.

The vertical diameter of this flat deposit was 1 mm, the horizontal 1,25 mm. It is situated 2,5 mm above the lower corneal border. To the right there are a few vessels in the deep parenchyma.

The picture was taken in March 1919, after the deposit had existed for three months, during which time it presented a practically similar appearance and form. The meshes originally were somewhat more dense and the colours more vivid. Evidently we are dealing with a fibrinous change. I cannot explain the cause of the colouration.

Fig. 56. Superficial remnants of obliterated corneal vessels.

Oc. 2, Obj. a2. Artist W. A. age 36. Limeburn of the cornea, one year ago.

Within the superficial parenchymatous layer there are numerous white dots and a few small grains of sand, the latter especially in the limbus. The white dots could not be removed by abrading the epithelium. Near the limbus there are some superficially situated straight vessel remnants, mostly branching dichotomously. They contain no blood, are white, and in certain places are directed almost parallel to the limbus. Often one may observe two parallel vessels, evidently artery and vein. In this case, and as a rule they are about 0,06 mm apart. The larger vessels are 0,02 to 0,03 mm wide.

Fig. 57. Delicate vessel remnants in quiescent parenchymatous keratitis.

Oc. 4, Obj. a2. Miss K. age 17. Parenchymatous keratitis due to hereditary lues. Right eye. Vision = 1.

This very benign attack of keratitis left deep vascular remnants which when observed in direct light (right part of figure) at first glance reminded of nerve fibres. They are of like diameters, however more tortuous and do not show the typic branching of nerve fibres. These vessel remnants are also whiter than nerve fibres. They are situated in the deep parenchyma.

By transillumination (right part of figure) one may observe the vessel remnants. They appear as *dark lines* mostly arranged in pairs (as a rule nerves are not visible by transillumination). The ends of the parallel vessel pairs are usually seen to join and form a loop. The length of the vessels averages 1,5 mm.

Fig. 58. Vessel remnants, deeply situated, many years after parenchymatous keratitis.

Mr. E. H. age 47. Seven years ago, for a duration of several months, he suffered a bilateral parenchymatous keratitis, which in form and type of attack resembled that due to hereditary lues. Today the Wasserman is negative. The eyes are free of irritation, and show faint diffuse cloudings in the deep parenchyma.

In the lower cornea, especially near Descemet's membrane, one may note straight vessels, mostly arranged *in pairs*, which are dichotomously branched. In direct light one may observe grayish-white bands, from 0,04 to 0,08 mm, a few as much as 0,12 mm in width. By transillumination these same bands are each seen to contain two vessels, an artery and a vein. Of these two vessels the wider one shows a double outline. The two vessels along their course separate in places and again approach one another. The endothelial reflecting zone (not shown in the illustration), presents numerous pitlike depressions, projected posteriorly.

The following illustrations in part present new observations I have made in cases of keratoconus. (Compare also Fig. 29.)

Fig. 59 and 60. Keratoconus. M. J. age 20.

The right eye shows a high degree of keratoconus, the left a very incipient one.

Fig. 59 presents the apex of the cone of the left eye under a low magnification. (Oc. 2, Obj. a2.) Fig. 60, the same under a higher magnification in focal light. To the right in both illustrations are shown the endothelial reflecting zones. Fig. 59 presents striped and streaked clumps of intensely white keratoconus opacification situated in the middle and deep parenchyma. *(Elschnig[122]), Strebel and Steiger[38]).)* Note the zig-zag direction and branching of this opacity. Behind this the vertical, in this case quite deeply situated keratoconus lines*.

They are composed of parallel dense gray lines or stripes, the majority of which present pointed ends, and as a rule they extend in a vertical direction.

According to the direction of the light (angle of incidence), these keratoconus-lines may present darker or again more luminous sides.

They remind one of the lines within the area of a traumatic annular corneal opacity (Fig. 52). At times one may note diagonal connections between the lines such as are seen when wood fibres connect a partially split board.

The *nerve fibre designing* is rendered distinct to a high degree. In a lower zone a fibre ends in a triangular white area, which latter measures about 30 microns. At its side is seen a fine white dot connecting with a recurrent nerve fibre. The cornea, especially the area at the apex of the cone was quite insensitive. The peripheral corneal area was less sensitive than that of the left less involved eye (*Arenfeld[36]*)). The endothelium was normal. We noted in this case, as well as in some other cases of keratoconus, that the patients had well worn flat incisors, exposing the dentine.

Fig. 61. Keratoconus. Mrs. G. age 38.

Focal illumination. Fig. 61 shows the changes at the corneal apex under a magnification of a medium degree. The white opacities of the superficial and middle parenchyma were more extensive and less distinct in outline in this case compared to the preceding one. In one area there is a quite dense opacity. The keratoconus-lines are less numerous and are deeply situated. The nerve fibre design is distinct. There are many irregular, often angular, areas[38]), at the fibre ends, or situated along their course.

At times, especially in older individuals, I have found "small white bodies" uniform in shape, within normal corneas.

In some instances they were found in connection with nerve fibres.

Therefore they need not necessarily be considered pathognomonic, in agreement with the findings of *Strebel* and *Steiger*.

At times a nerve fibres seems to have changed and several dots are seen along its course.

---

* These keratoconus lines have been erroneously reported by other investigators[37]) as folds in Descemet's membrane. Formerly we also presumed them to be such. Careful investigation of a large number of cases has taught us that they are not folds[28]). These lines may extend in any direction and are situated at varied depths in the parenchyma. These changes in the following text will be designated as *keratoconus-lines* in contradistinction to the *keratoconus-opacities*.

Fig. 55—62.   Tafel 6.

59

62

60

55

61

57

58

56

Vogt, Atlas.   Verlag von Julius Springer, Berlin.

In one place to the right and above a small dense spot of this type may be seen in the angle of a branching fibre. (The apex of the cone is quite anesthetic.) Below one may observe round opacities of the parenchyma, arranged in a row-like string of pearls. Just above this row of opacities a nerve fibre appears to be split for a short distance only, into numerous fibrillae.

Fig. 62. Keratoconus. Dr. J. G. age 26.

Apex of the right keratoconus. The left cornea presents an incipient conus development. Focal illumination. The *keratoconus lines* in the deep parenchyma are mostly vertically arranged. The endothelial cell design is somewhat indistinct, in parts amorphous. Below, there are three faint incipient lineal *opacities*, one of these is zig-zag in shape. The nerve fibre designing is more distinct than is seen in normal corneas. To the right and below a nerve fibre branch shows an abnormal thickening and increased density. (All visible nerve fibres are not seen in the illustration.) The apex of the cone is somewhat anesthetic.

Figs. 63 and 64. Keratoconus of the right (Fig. 63) and left (Fig. 64) eyes of Mr. K. age 20.

He is unusually tall and anemic. The keratoconus is supposed to have developed within the last year. It is well advanced in both eyes. The nerve fibre designing which is especially distinct is not shown. (With the exception of a few fibres in Fig. 64.) The opacification of the apex of the cone is quite intense. It is confluent and forms a compact opaque area. In certain places it is especially dense and striped. Within this area, anterior and posterior to it, *at varied depths*, there are numerous *keratoconus lines*.

In places they form a delicate lattice-work, the cross fibres however are at varied depths, some anterior and others posterior to the keratoconus opacity, as well as to the nerve fibres. To the left and above, the lines diverge to form a flame-like spray.

As a rule the direction of the lines is a vertical one, at times oblique or horizontal. I especially noted vertical lines in cases which presented an astigmatism against the rule.

The right cornea showed (at first only nasalward) a broad rupture in Descemet's membrane, to be seen in Fig. 63. Its edges are seen distinctly in indirect light only (transillumination), while the balance of the illustration presents the change in focal light (compare Fig. 66).

The upper ruptured edge is rolled up, which causes two reflex lines to be seen within it by transillumination. The widest separation of the edges of the ruptured membrane is about 0,5 mm. The separation between the two reflex lines, due to rolling up of the edge measures 0,04 mm. The double lines are continuous in the opaque area in the form of a strung pearl-like row of brilliant dots, within round, darker areas. The nature of the latter is obscure. Three weeks later a rupture of a similar type was discovered temporalward.

In connection with these ruptured Descemet's membranes we noted a *considerable thickening* of the apex of the cornea, easily revealed by the slitlamp. It is probably due to an absorption of aqueous by the parenchyma.

The examination of the blood of this patient by *Dr. Löffler* of the medical clinic of *Prof. R. Staehelin*, showed no abnormality. His incisors were flattened as if it had been done with a file.

The described rupture of Descemet's membrane can only be ascribed to the development of the keratoconus. There is no hydrophthalmos, and a trauma at the time of birth is excluded by the history.

The fact that the greatest separation of the edges is situated at the apex of the cornea proves its appearance is due to the distention accompanying keratoconus. In addition we have the antero-posterior thickening of the right apex, which is not present in the left eye.

These ruptures of Descemet's membrane in cases of Keratoconus were first described by *Axenfeld*[39]). He also noted the thickening of the parenchyma at the apices of the cones and pronounced them a result of the ruptures. (Compare also *Unthoff*[124]).

Fig. 65. Nerve fibres visible by transillumination in the case of keratoconus of Fig. 63 (temporal and downward).

The well developed nerve fibre designing is also visible by transillumination in this case. The nerve fibres appear as stripes, for they and their immediate surroundings reflect the light. The side of the fibre toward the light is dark, the opposite one is luminous. That these stripes are nerve fibres can be proven by alternately examining them in direct light and by transillumination.

Fig. 66. The nasal and temporal rupture of Descemet's membrane in the case of Keratoconus, shown in Fig. 63.

Oc. 2, Obj. a2. Transillumination. Delicate dewlike changes of the endothelium in a large area surrounding the rupture. Three months ago the upper edge showed a double reflex line, (rolling up), now the lower edge also presents the same change. The nasal rupture is observed with the light temporalward, the temporal one with the light on the nasal side.

Fig. 67. The posterior reflecting zone in the area of the temporal end of the ruptured Descemet's membrane in the preceding case.

Oc. 2, Obj. a3. The endothelial reflecting zone is interrupted in the area of the rupture. The edge which forms the opening extends at an obtuse or a sharp angle, according to the relation of the direction of the angle of observation, to that of the light. (Compare page 23.) A double reflex line showing a parallactic displacement, which can be made to wander at will, proves the presence of an irregularity of the posterior corneal surface within the area of the rupture.

The *irregularity* of the posterior corneal surface in the area of the rupture is shown in the reflecting zone.

Fig. 68. Megalo-cornea with a ruptured Descemet's membrane. (Axenfeld[39]*), W. M. age 16.*

His left eye was lost in infancy on account of hydrophthalmos. Right eye, lower half of cornea. Oc. 2, Obj. F55. Focal illumination. Horizontal diameter of the cornea 13 mm. The largest rupture was 0,8 mm in length, others 0,6 and 0,7 mm. The width of the border which showed a double reflex line is 0,04 mm. The direction of the rupture is mainly one concentric to the limbus. These ruptures, as is known, are the cause of the band-shaped opacities of *Haab*[40)][41]), in cases of hydrophthalmos. Subsequently the opacities clear up. The ruptures however remain visible.

(Compare also the anatomical investigations of *Reis, Seefelder* and *Staehli*.)

Fig. **63—73.**   Tafel **7.**

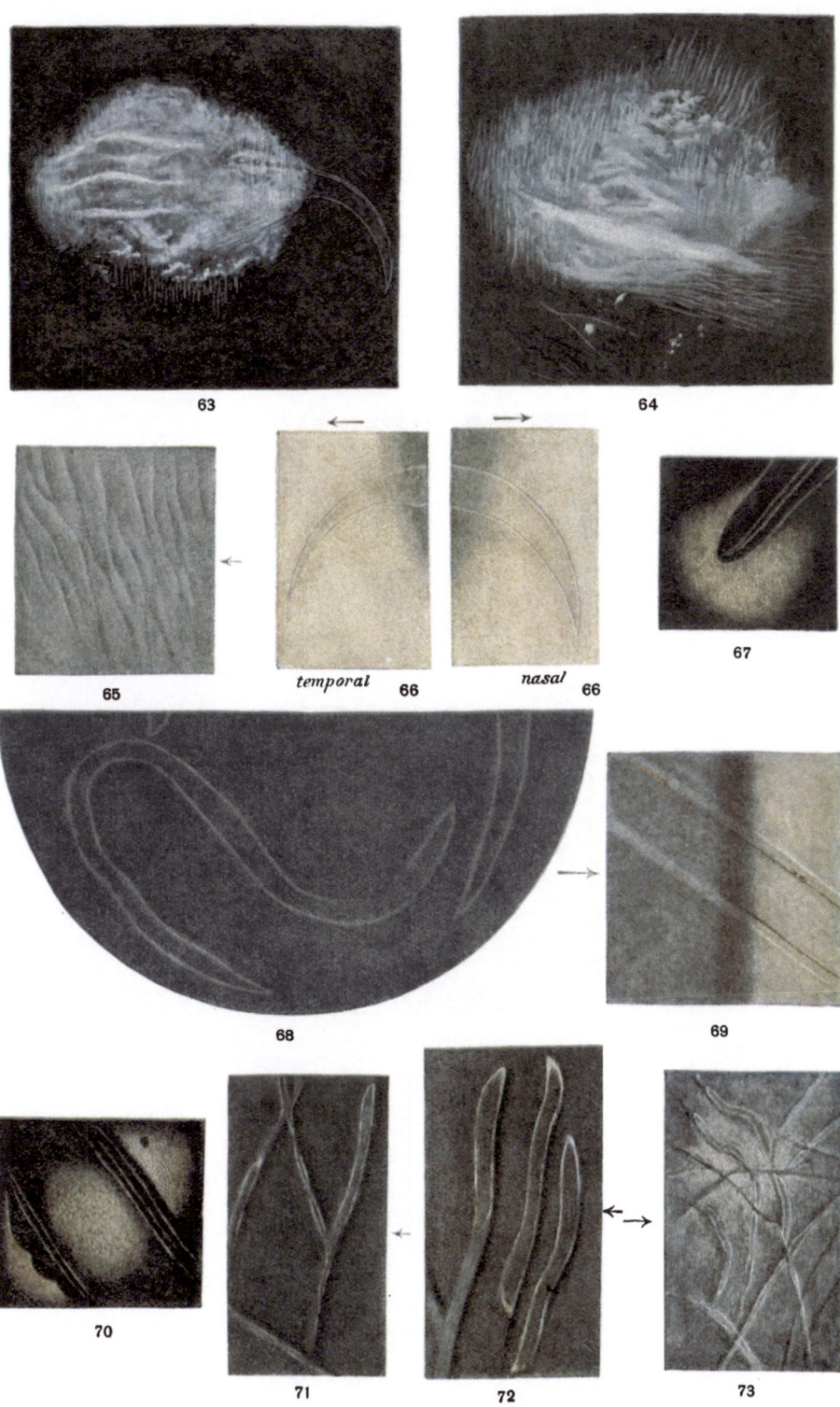

63

64

65

temporal   66   *nasal*   66

67

68

69

70

71

72

73

Fig. 69. An area of ruptures in Descemet's membrane in the same case under a higher magnification.

Oc. 2, Obj. a2 seen in direct light (left) and by transillumination (right).

The double reflection is more distinct by transillumination (to the right). It is due to the rolling up of the ruptured edge of the membrane. By the chosen angle of observation the outer edge of the rolled up border is luminous and the inner edge is dark. Deposited pigment is black.

In contrast to this by direct light (to the left in the illustration) the double lines are hardly visible. The tissue strip between them however is *luminous on a dark background*, and the pigment deposited on the edge of the ruptured membrane is seen in its natural red color.

Fig. 70. Posterior reflecting zone in the area of the ruptured Descemet's membrane.

Case of Fig. 68. Oc. 2, Obj. a3. An area near the temporal lower limbus is shown. Near the ruptured edge there are two or three reflex lines, more or less visible according to the relation of the angle of observation with that of the direction of light. These lines, due to the rolling up of the edges, are separated from the endothelial zone by a dark interval. To the temporal side this interval presents waving variations in its width, while to the right it is of equal width throughout its entire length. This wavy variation is not necessarily due to endothelial defects, but is in all probability caused by an irregularity of the surface. To the right and upward in the endothelial zone there is a round pit-like excavation extending posteriorly, which resembles a hole. While the endothelium surrounding the ruptured areas is normal in appearance, the area between the two edges shows an amorphous designing and a distinct *iridescence*. The irregularity of the posterior corneal surface, as in the case of Fi·. 67, is here also shown in the reflecting zone.

Figs. 71 to 82. Reflex lines due to folds in Descemet's membrane.

Regarding the genesis of these reflex lines compare page 23 and the text pertaining to the following illustrations. That these deeply situated corneal stripes were folds in Descemet's membrane was in all probability first noted by *Albrecht v. Graefe*[129]). Regarding experimental proof see *Hess* and the author[28]).

Figs. 71 and 72. Folds in Descemet's membrane in the case of Mrs. B. age 66.

She was operated on for cataract seven days ago. Folds of this nature are regularly found shortly after cataract extractions. They are grayish lineal stripes at right angles to the direction of the incision. For a long time they have been known as "striped corneal opacity"[42])[43]), (formerly keratitis striata).

These characteristic reflex lines have not been seen by former observers. (However, compare an observation by *Dimmer*[44].) Double reflex lines of this type according to our observations are characteristic of folds in Descemet's membrane. Within the lines the endothelium is usually amorphous in character. The lines are regularly arranged in pairs. The one line corresponds to the convex, the other to the concave cylindric reflection of the folded membrane. Toward their ends the two converge to form one line. According to the angle of observation and the radius of curvature of the folds the lines may appear more or less apart or flattened. At times there may be constrictions of the pairs (Figs. 71 and 73), at which places the lines are seen to approach one another. Interruptions may also be noted, which appear and disappear by changing the angle of observation.

To study the origin of these reflex lines freshly enucleated human, pig and calves eyes may be used (Figs. 83 to 90).

By the aid of plastic models made of ribbed black enamelled glass or of black paper covered with collodion, the genesis of these lines may be easily studied. They always originate on the sloping curve of a fold and are therefore not in any relation to the width of the latter. In human eyes we have not alone observed these reflex lines in folds of the posterior limiting membrane of the cornea, but also in Bowman's membrane, the conjunctival surface, the lens capsule (in shrinking of the lens or in secondary cataract)[28]), as well as at the border of the retina and vitreous. In the latter cases they are due to folds of the superficial retinal layers.

These lines, visible in the cornea, are of importance in diagnosis. They prove the presence of folds of the posterior corneal surface in cases of traumatic and nontraumatic perforation (Figs. 71 to 82), and when Descemet's membrane is contracted by scar formation. They are almost found with regularity in all cases of parenchymatous keratitis, for instance in discform keratitis, while they are but rarely seen in superficial keratitis.

In ordinary illumination these reflex lines are often not distinctly seen, the folds only show as gray stripes, hence the former designation "keratitis striata".

According to the rules we have established, one may vary the *distance* which separates these reflex lines due to folds, by changing the relation of the angle of observation and that of illumination. We may increase the angle of observation by alternately observing the lines through the right and left eyepiece. This enables us to observe changes in their separation by noting, respectively their appearance and disappearance, proving its dependence on changes in the angle of observation. Regarding the increase in the width of these double reflex lines, the causes of their convergence, segmentation and the relation of the shaded lines to the latter, compare the observations of the author pertaining to this[28]).

Fig. 73. Irregular folds in Descemet's membrane following extraction with conjunctival flap. Miss E. D. age 70.

Extraction of senile cataract ten days ago. Oc. 2, Obj. a2. In addition to folds in their usual direction, one may note in a temporal area a whorl-like centre from which vertical, horizontal and diagonal folds are seen to radiate. In the centre of this whorl the reflecting zone shows yellow amorphous endothelium. Several folds present a segmentation of the reflex lines, in parts they are single. In the area of the reflecting zone two vertical folds are crossed by a "shaded line" (see below), which runs in a horizontal direction.

Fig. 74. Irregular folds in Descemet's membrane and a superficial corneal fold, near an extensive traumatic perforation wound. Farmer J. B. age 62.

Oc. 4, Obj. a2. Eight days ago he sustained an extensive corneal perforation by a cow's horn. Fig. 74 presents a part of the lower external cornea. The straight, white double line is due to a lineal kinking of the anterior corneal surface. The posterior corneal surface is thrown into irregular curved folds, the main direction of which however is one parallel to the direction of the kinking. Again observe in this case the not always easily discernible shaded stripe along the double reflex lines. Anatomical examination confirmed the presence of the folds.

Fig. 75. Parallel dense folds of Descemet's membrane following extensive corneal perforation. Master F. age 10.

Oc. 4, Obj. a2. Injured by a stone which hit his eye two weeks ago. Corneal perforation and iris prolapse. The anterior chamber in certain areas has not re-formed. The dense folds are quite close to one another and are due to a relaxation of the posterior corneal surface. To the right a single fold may be seen running in a different direction. The surface of the folds is *dull* in certain places and the reflex lines are not very distinct. The reflex zone shows an amorphous yellowish granulation.

Fig. 76. Indistinct folds in Descemet's membrane due to cicatricial contraction Mr. H. V. age 38. Perforation at the nasal limbus.

Oc. 2, Obj. a2. Focal illumination. The injury occurred five months ago. The width of the yellowish reflex lines shows that the folds are flattened. The deep corneal parenchyma presents a criss-cross latticework of dark lines (shadow lines), as were described in Fig. 79. Note that these dark lines in places *cut through* the reflex lines. This proves that both are within the *same* optical plane. (Regarding these also compare Figs. 73 and 80 and the microphotographs of Figs. 84 to 89.) We therefore have folds of Descemet's membrane crossing one another in which the reflex lines are visible only in certain parts, according to the angle of illumination.

Figs. 77 and 78. Dull folds in Descemet's membrane.

If the cornea is opaque in the area of Descemet's membrane, for instance after severe parenchymatous keratitis or extensive diffuse deposits on the posterior corneal surface in iridocyclitis, a reflection of the posterior corneal surface is impossible, nor can we demonstrate the endothelial reflecting zone. Every fold in this case will therefore show a luminous and a dark area. The luminous area reflects diffuse light. If one wishes to compare the reflections of dull and luminous folds, folded blotting paper and glazed paper may be utilized.

Master E. Sch. age 15, suffered a bilateral attack of parenchymatous keratitis due to heredetary lues, five and four years ago. In both eyes the deep parenchyma is opaque and presents many parallel vertical slightly curved folds (compare also the observations of *Dimmer* [44]).

Fig. 77 presents the left cornea under a low magnification. On the posterior corneal surface on the quite opaque membrane of Descemet, there is deposited a large quantity of brown pigment. The double reflex lines are faintly visible in a few places. Just anterior to the folds, in a uniformly deep parenchymatous layer, and at right angles to them one may observe straight vessels which are dichotomously branched.

The bright surfaces of the folds are separated 0,05 to 0,1 mm from one another. The areas between the folds are relatively flat and are at least again as wide as the folds. The vessels are about 20 mm in diameter.

Fig. 78 presents the folds of the right eye under a higher magnification (24 times). Oc. 2, Obj. a2.

The folds are probably caused by a retrogression of the extensive corneal oedema in parenchymatous keratitis and comparable to the wrinkling of the skin of a dried apple. Owing to the corneal macula the vision is reduced to $^6/_{24}$ in the right and $^6/_{36}$ in the left eye. Anterior astigmatism not worthy of mention.

Fig. 79. Endothelial reflecting zone and dark lines in Descemet s membrane (shadow lines). Mr. Z. age 75. Ten days after lineal extraction of a traumatic cataract.

Oc. 4, Obj. a2. In focal light.

The endothelium is yellowish, amorphous, and granular, the reflecting zone and its surroundings are criss-crossed by the shadow lines seen in Figs. 73 and 76. The endothelial zone ends abruptly at the lines, proving that the latter are due to an interruption of the reflection by the folds. Over some of the very faint lines one may observe the continuation of the endothelial surface, proving that the folding of the membrane at this place is not sufficient to cause a complete disappearance of the reflection that is producing a dark line. On the left side one may observe a shadow line having on its one side a slightly luminous edge. This edge also proves the presence of a change in the surface curvature. In well developed cases, with the proper angle of observation, one may see the parallel pair of reflex lines adjacent to the shadow line (compare Fig. 72). In other words the shadow line corresponds to that part of the fold which does not reflect light into our observing eye. The reflex lines in contrast are due to the reflection of the concave and convex cylindrically curved surface of the folds[28]).

Fig. 80. Irregular folds in Descemet's membrane due to a cicatrix in the posterior corneal surface.

Oc. 4, Obj. a2. Mr. S. M. age 59, had a preliminary iridectomy performed on his left eye 5 years ago. In withdrawing the keratome the operátor injured the posterior corneal surface. An opacity of the parenchyma developed, which today with indirect illumination shows pigment debris and dewlike changes. Radiating from this scar there are irregular folds in Descemet's membrane, showing the reflex lines as are illustrated in Fig. 80. Below these there is an extensive yellowish reflecting zone with indistinct endothelial cell borders. Above, the edge of a reflex line is lost in a flat endothelial reflecting surface. The scar is so white and dense that it fails to reflect the posterior corneal surface. At the lower border of this cicatrix there is a small brown pigment deposit.

Fig. 81. Simultaneous folds in Descemet's and Bowman s membranes.

Mr. H. age 24, injured two years ago by the explosion of a mine. Low magnification. No irritation. Tension slightly reduced. The white circumscribed spots are particles of stone imbedded in the superficial corneal parenchyma. A dense perforation scar may be seen in the nasal limbus. (To the left in the illustration.) Above this scar there is an extensive bulging of the iris (iris cyst?). The folds in Bowman's membrane show as indistinct reflex lines, in parts resembling single opaque stripes. In the illustration there are white lines' visible which connect the imbedded particles of stone. In certain areas one may quite distinctly observe the double reflex lines. More daintily graceful are the white lines of the folds of Descemet's membrane, which latter are evidently due to traction, caused by the cicatricial contraction. In contrast to the folds which follow cataract extraction and are mostly surrounded by a slight opacification of the tissues, the latter in this case are clear in the vicinity of the folds (compare text to Figs. 71 and 72).

Fig. 74—83.   Tafel 8.

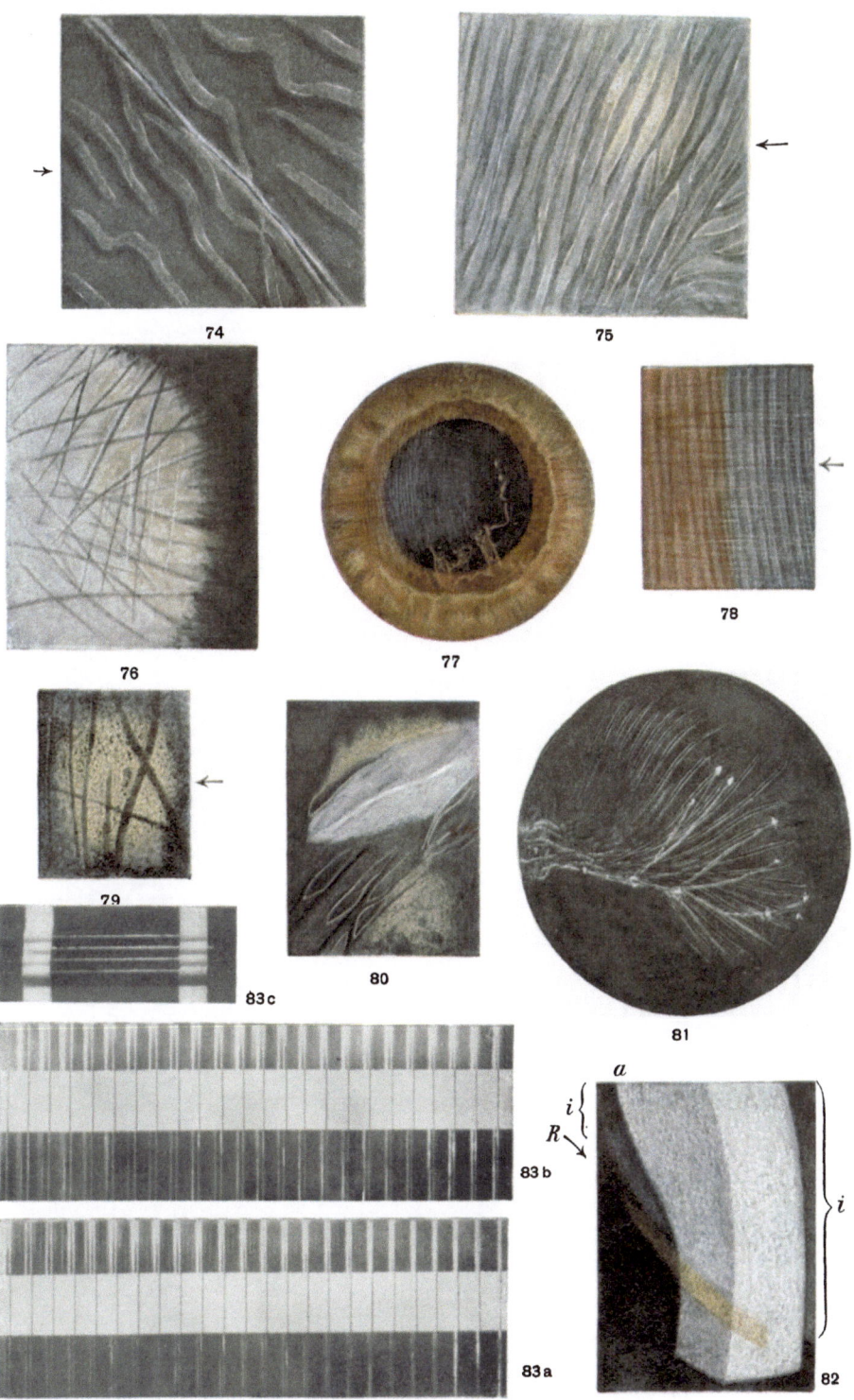

Fig. 82. Schematic representation of the annular reflexes of the posterior corneal surface in discform keratitis.

These annular reflexes were produced by injecting water by means of a delicate needle into the deep parenchyma in freshly enucleated eyes of mammals. This not alone causes an opacity but also a belly-like swelling of the parenchyma extending into the anterior chamber (Fig. 82), easily visible by slitlamp transillumination. This prominence is surrounded by a ring-formed reflex line, caused by the circular reflection in its periphery. As the reflecting surface is convex anteriorly, the ring reflex is posterior to it in the depth of the anterior chamber. It shows by focussing the *reflecting zone* of the endothelium. I produced this annular ring reflex in a most distinct manner by injecting water into a circumscribed area in the corneas of living rabbits. A bulging resulted, at times anteriorly, at times posteriorly. Surrounding this bulging in a circular manner (Fig. 82) one may note in the direction of the zone of diffused light the beautiful bronze-like luminous reflex ring, which seemed to be suspended in the anterior chamber. By focussing the reflecting zone which produces this reflex, that is the periphery of the prominence, one may observe the endothelial cells with their hexagonal borders (*i, i'* the area injected with water). The same and very similar annular reflexes I have observed surrounding circumscribed parenchymatous keratitis. These latter forms (for instance, discform keratitis) always produce a thickening of the cornea in the direction of the anterior chamber, visible by slitlamp illumination. This thickening is surrounded by the annular reflex which seems suspended in the aqueous. The slitlamp is an excellent help in assisting to reveal circumscribed and diffuse thickenings of the corneal parenchyma. Not alone a thickening but also a thinning of the cornea is discernible, for instance in ulceration (ulcus serpens etc.), threatened perforation, and in keratoconus, as well as in old scars, by the use of the slitlamp. (Narrow slit.)·

Formerly we were clinically unable to diagnose these changes in thickness of the cornea. As the bundle of light does not enter the area under examination at a right angle, mistakes may be made by a diagonal direction of observation. It would be an advantage if the slitlamp were so constructed as to allow a rotation of 90° around its longtitudinal axis.

Figs. 83 to 90. Photographs and microphotographs to illustrate the double reflex lines caused by folds in reflecting limiting surfaces. (Compare page 21.)

Fig. 83. Black enamelled ribbed glass plate. The radius of curvature of the ridges is quite less than that of the depressions. The latter radii diminish in length as we approach the border of the concave reflection. The vertical lines of the horizontal paper stripes are the summits of the ridges. In addition to the double reflex lines one may observe in well illuminated areas, the less luminous secondary lines. The distance of the lines increases toward the left (a diminution of $\frac{\varepsilon + \beta}{2}$), *a* and *b* show the lines to the right in its larger value, to the left in the reduced value of $\frac{\varepsilon + \beta}{2}$.

Note the increased width of the stripe of the depression due to the increased radius of curvature, which in this case is especially pronounced on account of the reduced value of $\frac{\varepsilon + \beta}{2}$ (compare illustration *b* to the left).

Fig. 84. Folds in ·Descemet's membrane in a man 60 years of age, seen from the front. In certain areas the reflexes are flat. Confluence of the ends in parts show an increase in the width of the reflex lines. (Owing to the irregularity of the cornea but few lines are distinctly seen at one focal point.)

Fig. 85. Reflex lines due to traction of Descemet's membrane in a calf's eye, viewed from the rear. Parallel quite closely arranged narrow long forms, having pointed ends.

Fig. 86. Branching of folds (Descemet's membrane of a calf) in parts reflex lines of the reduced value of $\frac{\varepsilon + \beta}{2}$. Folds due to a rupture, split upward.

Fig. 87. Irregular in parts flat reflexes curved in various directions, due to folds in Descemet's membrane in a pig (viewed from the rear).

Fig. 88. Crossed folds in the anterior corneal area (pig).

Fig. 89. Folds in Descemet's membrane of a calf, showing a segmentation.

Fig. 90. Reflex lines due to folds in the bulbar conjunctiva of a three months old child.

Fig. **84—89.**　　　　　　　　　　　　　　　　　　　　Tafel **9.**

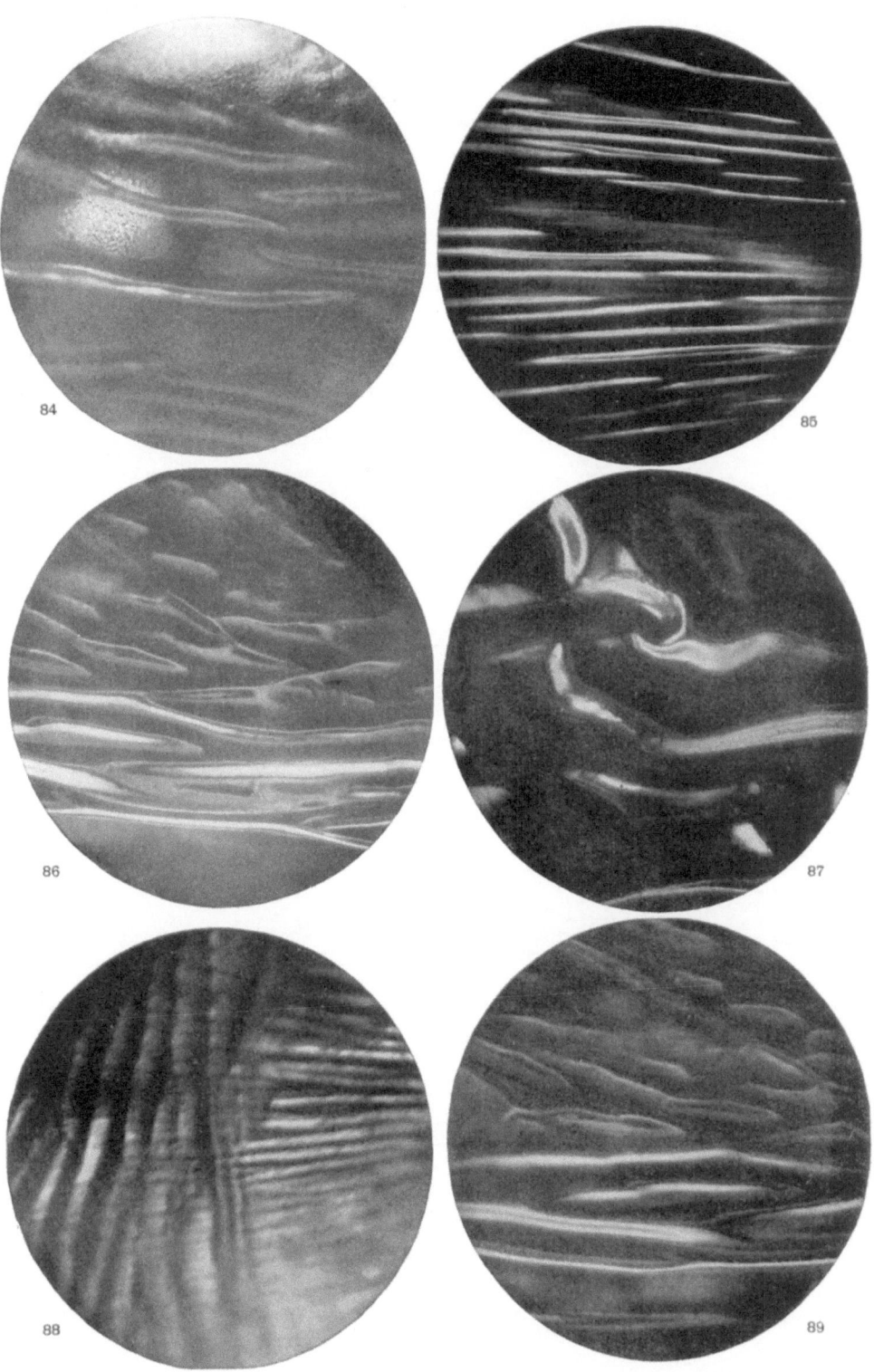

Fig. 84—90.

EXAMINATION OF THE LENS

(Wherever it is not especially so stated, observations are made in focal light)

1. THE NORMAL LENS

DETERMINATION OF DEPTH IN THE LENS

The estimation of depth in the lens with the slitlamp has not as yet been described, stereoscopic observations were considered sufficient. Exact measurements are possible only if we utilize the intensely luminous shaft of light in connection with the areas of reflection.

The slitlamp will not alone be of aid to the general ophthalmologist, but will disclose new paths for investigation to the research worker in the field of cataract. The nitrogen lamp is also to be preferred, in the described modification *(focussing of the fibre onto the diaphragm of the illuminating lens)* unless one is observing very minute colour variations.

The principles guiding the determination of depth in the lens are similar to the ones which have been given for the cornea (see page 8). The slit in this case also must be *reduced* in width to $^1/_2$ mm or less. (The width of the focal bundle should be from 0,05 to 0,1 mm. In the illustrations a greater width was chosen.)

The anterior edges ac and bd which are the lateral border lines of the *anterior capsular stripe* (Fig. 1a) when viewing the anterior lens surface, are not easily discernible in *diffusely* reflected light unless this surface has lost transparency by the deposit of exudates, or from other causes.

However it is possible to observe this border line in all cases by the use of the bundle of light in focal illumination. We now have this exact method which for the first time allows us to see the anterior capsule and which no doubt will be of great aid in future research work, pertaining to cataracts.

By focussing the *anterior lens reflecting zone* a sharper definition of edges ac and bd is obtained (see page 20). As the reflecting zone of any specific part of the anterior lens surface may be focussed with little practice (compare Fig. 93), therefore these two so important edges necessary for the determination of depth are easily located.

One must remember that every point which is seen by movement of the lateral border of the bundle of light on the side of the objective (surface $bdfh$ Fig. 1a in which bd is in the anterior capsule and fh in the anterior surface of the senile nucleus), is definitely localized by virtue of its appearance and disappearance.

This lateral border of the bundle of light is, if in good focus, sharply demarkated on the *anterior capsular surface.* (The surface of entrance $abcd$ Fig. 1a.) It is thus defined as a plane, which we must conceive as being situated between bd and fh.

For the purpose of accurate observation, for instance, for the localising of lineal and flat opacifications, I recommend the use of the 1 mm round diaphragm (see page 10), which is found on the rotating disc with the slit diaphragm. I again mention that the angle between the direction of observation and illumination must not be acute. When using this method of the "cylindrical bundle", it is best to avoid the reflecting area of the anterior surface of the lens.

The *lamellar surfaces* in the lens substance are also visible. They are somewhat dull (matt) and yellowish-red in colour. As the curvature of the anterior surface of the nucleus of the senile lens is greater than that of the anterior surface of the lens, it may not always be possible to simultaneously observe the reflecting zone of these two surfaces, within one and the same angle of observation. If, for example, with the illumination temporalward, we have the reflecting zone of a temporal lens surface area in focus, it will be necessary to have the patient rotate his eye slightly to the temporal side, in order to be able to see the reflecting zone of the senile nuclear surface.

One may thus make the surprising discovery that in relatively young persons the nuclear surface line is more luminous than that of the superficial lens surface, when both are simultaneously seen. This is due to the fact that now the nuclear surface alone presents the reflecting zone.

In the case of binocular observation we may see, with one ocular, the reflecting zone of the lens surface, with the other, that of the nuclear surface (see Fig. 91a).

$L =$ entering light, $Ch =$ light reflected from the anterior lens surface, $N =$ light reflected from the anterior surface of the senile nucleus.

An eye observing in the direction Ch would only see the anterior graining (shagreen), one observing in the direction N, only the reflecting zone of the anterior nuclear surface.

After the beginner has learned to focus the anterior capsular stripe with and without the reflecting zone, he may alternately direct the bundle of light onto the anterior capsular and anterior senile nuclear surfaces. He may thereby determine the *depth of the cortex*, and the increase in its depth from the axial to the peripheral areas. To do this he must alternately focus the capsular and senile nuclear reflecting zones.

Furthermore, he may determine the location of various *cortical opacities*, if such are present, by noting the site of their appearance in relation to bd and fh when the bundle of light is moved from side to side. The principles for determining this are given on page 8. The beginner must remember that a point if seen in the *direction* of the anterior capsular stripe need not necessarily be situated within it.

The *change of position* of the bundle of light in the manner as described above alone determines its location. After using the narrow slit bundle it is advisable to verify detail with the aid of the "cylindrical bundle".

It is therefore evident from what has been stated, and from the observation of Fig. 91a in which manner the lamellar surfaces can be utilized for the precise determination of depth, in focal light.

It need not be further emphasized that the relative position of *circumscribed infiltrations*, unless slight differences of depth (niveau) are present, may be recognized by stereoscopic observation.

For the localization of the *capsule, the lamellae, flat infiltrations* covering a large area, the borders of cystoid spaces, spokes etc., we have only the method of utilizing the bundle of light, the use of which however necessitates some experience.

It need not be again mentioned that in this connection, stereoscopic observation is of aid in simplifying matters. The intensity of the zone of light within the lens is somewhat less because of its reduced internal reflection and fluorescence, in comparison to that of the cornea.

For a practical example of the determination of depth in the anterior cortex see Fig. 212 also the text to Fig. 277 a and b.

If one wishes to observe a lens opacity by *transillumination* with the slitlamp, I would recommend the use of the reflecting zone, or the reflected image of the posterior lens capsule, providing the lens substance posterior to the opacity is sufficiently transparent to allow this.

Fig. 91a and b, see text regarding the determination of depth, and on page 5.

Fig. 91c. Lens of an unusual depth. Woman age 70 yrs.

Glaucoma, quiescent after iridectomy and trephining, Cataracta incipiens.

Fig. 91d. Lens of a medium depth. (Mr. Sch. age 40 yrs.)

Fig. 92. The normal anterior reflecting zone of the lens Fridenberg[144]), Tscher-ning[143]) u. a. (anterior lens graining or shagreen C. Hess[45])). Patient age 18.

Fig. 91 = Oc. 2, Obj. a3; Fig. 92b = Oc. 2, Obj. a55.

The reflecting zone of the anterior surface of the lens is large because of the radius of curvature of that surface. (Compare with it the smaller zone of the posterior lens surface Figs. 96 and 98*.)

The epithelium of the lens is not so easily seen as the endothelium on the posterior corneal surface, and is only visible in small areas of the reflected zone. The individual epithelial cells are hardly discernible. One must not confound the coarse design seen by as low a magnification as is the corneal epithelium, with lens epithelium (see Fig. 92 and 93). The latter cells are much the smaller. The areas of the coarse designing just mentioned may at times appear as depressions or elevations.

At *N* there is a seam (suture) and adjacent to it the *surface of the fibre.* Especially in observing peripheral areas of graining, one will note an apparent parallelism and a formation of ridges in the direction of the fibres[46]).

In the central fields, the irregular designing, depressions and elevations predominate, as is shown by Figs. 93 and 94 in pigs' eyes.

The appearance of the dark seams (sutures), and fibre designing in the area of graining proves that not alone the epithelium, but also the superficial fibres and seams participate in its formation[46]). When using the nitrogen and arclight, I have even at times observed the superficial fibre system *outside* of the area showing the shagreen.

The borders of the epithelium of the lens is more difficult of observation compared to the corneal endothelium, because it is located under the intensely reflecting lens capsule. (The differences between the indices of refraction of the capsule, epithelium, and cortex, are less than that between capsule and aqueous?) The light reflected by the aqueous tends to veil the epithelial designing. The epithelial areas are more distinctly seen with the nitrogen and arclight. With these, under a magnification of 37—68 times the *cell borders* may be seen, within limited areas. The best illumination for this purpose is given by the "cylindrical bundle". (Page 10, the application of the principles of achromatic optics is to be recommended). Areas of epithelium on the lens capsule are *uniformly irregular,* when contrasted with the

---

* In animals the radius of curvature is less, hence the reflecting zones are smaller (see Fig. 98 and 94).

corneal endothelial surface. The epithelium is more distinct in youth. It is noticeable that the graining often abruptly ends at a seam (Fig. 92 to the left and up), or that the interposition of a seam may vary the designing, within the same focal field. By using a focal bundle area the shagreen shows as a *sharply circumscribed* stripe, as for instance in Fig. 93. In a bundle area which is not sharply demarkated it becomes faintly lost in its surroundings (Fig. 94).

For the method of focussing for the anterior lens graining, see under "Technic". If we have focussed a central area and we wish to observe one more nasal, we must direct the patient to look in the *reverse* direction.

The designing is only visible in the area of the reflecting zone. We can observe a reflecting (grained), and a diffusely reflecting zone on the lens surface, in a similar manner as we see it on the cornea. The zone of diffused light appears *dark-gray in normal eyes* (Fig. 93).

Figs. 93 and 94. Normal anterior graining of lens, (pig's eye) freshly enucleated and in situ.

The coarse fields are not individual cells. The latter are much smaller and visible only as fine dots.

Fig. 93 shows the zone with a lineal magnification of 24 times. Figs. 94, 68 times. Fig. 93 shows the strip of diffused reflection at "*D*", "*Sp*" shows reflecting surface zone.

Fig. 95. Micro-photograph of the anterior graining (pig's eye).

Magnification 16 times with the slitlamp. Compare text of Fig. 93 and 94. The individual cells are visible in some places as groups of small dots.

Figs. 96 and 97. The posterior reflecting zone of a normal lens.

Oc. 2, Obj. a2. *Fig. 96.* The posterior reflecting zone at the posterior pole also shows a graining. The fields are irregular, long and often wavy.

If we focus a peripheral area (Figs. 97 and 98) we note a fine fibrelike striping which is a product of the *lens fibre surface.* Nasalward and temporalward the direction of the fibres is a horizontal, correspondingly above and below, a vertical one.

The centre of the surface zone shows an irregular graining, the latter continues toward the periphery in the fibre-like design (Fig. 97), which radiates toward the clearly defined *sutures* of the posterior lens fibre surface (Fig. 98).

Deposits on the posterior lens capsule show an interruption in the reflection, in a similar manner as seen on the posterior surface of the cornea. They are *dark* on a *luminous background,* and are mostly remnants of the foetal membrana vasculosa— visible in Fig. 96 and 97 as small dark dots, spots and stripes. *Subcapsular* deposits are seen in the form of dark and coloured spots, dots, or small flat surfaces (see below), within the reflecting zone.

Surrounding the reflecting zone ("*Sp*" in Figs. 96 and 97) is the diffusely reflecting posterior lens surface "*D*", in which the changes appear *light* on a *darker background* (areas of greater reflection in focal light).

The technique is not difficult, one can easily find the yellowish luminous shaft of light on the rear lens surface, and then focus its reflecting zone. To shift the zone the patient will look in a direction corresponding to the displacement desired.

Fig. **90—99**.   Tafel **10**.

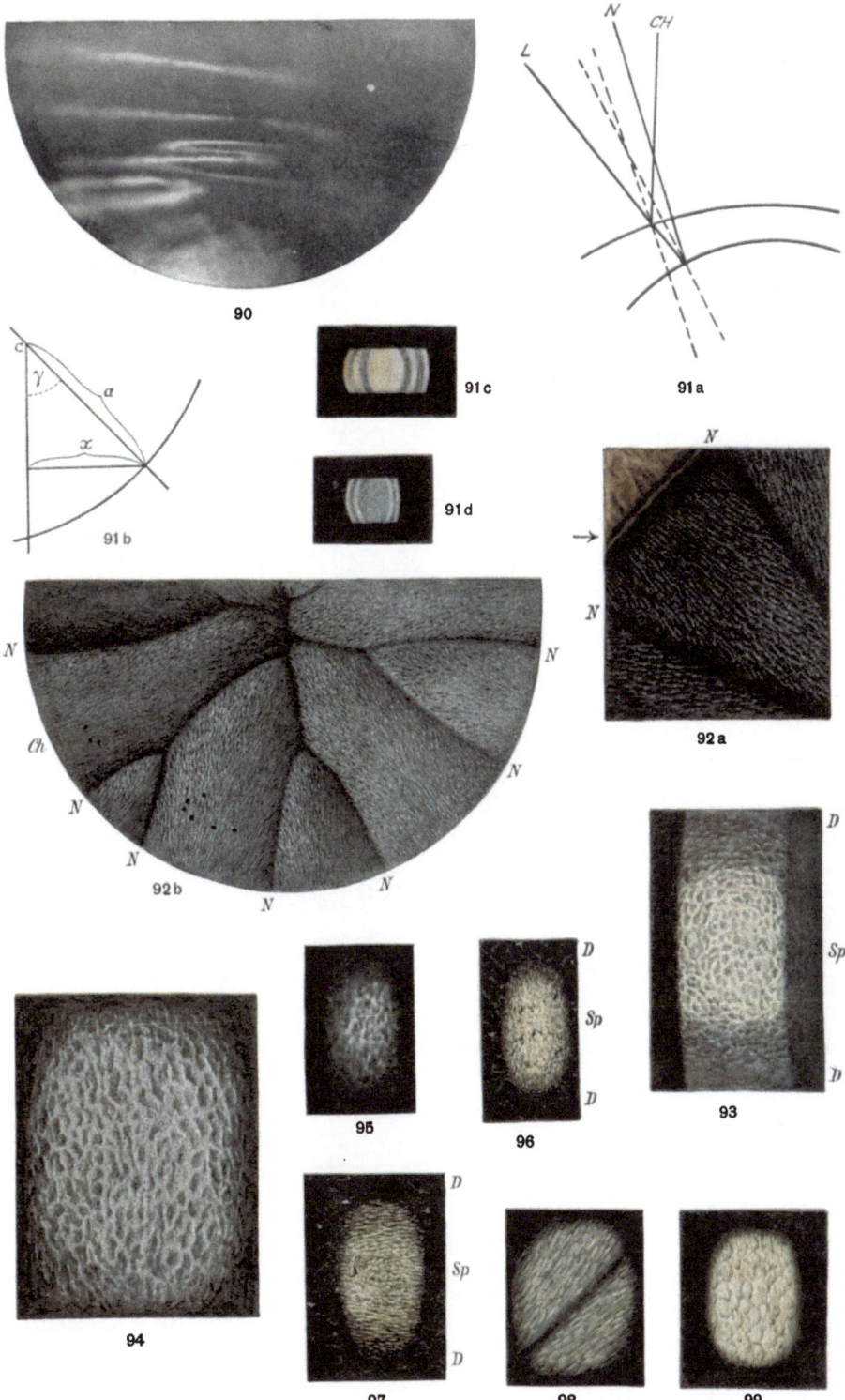

Vogt, Atlas.   Verlag von Julius Springer, Berlin.

Fig. 98. Posterior reflecting zone and sutures.

Oc. 2, Obj. a3. Male Schw. age 17, superior temporal lens area. The light stripes are the fibres which join the (dark) sutures.

Fig. 99. Mosaic-like fields on the posterior reflecting surface. (Instructor A. age 65. Advanced glaucoma simplex.

I have seen similar fields composed of white lines in other, in part, youthful eyes, however the lines were less distinctly seen. These fields are in the vicinity of the posterior lens pole. In the glaucoma case, I at first supposed them to be proliferations of the capsular epithelium. However the large size of the field (up to 60 microns) and my later observations in youthful eyes, has disproven this. Owing to their form these fields cannot be identical with the fibre impressions of the posterior capsule as described by *Henle*[47]), *Barabaschew*[48]) etc.

Fig. 100a. Sagittal meridional section of the lens. Sketch of the concentric zones.

The sagittal section is seen diagonally from the front. A quadrilateral prismatic bundle of light LL passes through the lens in the direction of the arrow and in this manner exposes six white ribbonlike luminous lamellar surfaces. Between these one may observe the dark intervals. The two middle white bands show the Y shaped embryonic sutures.

This illustration is for the purpose of guarding the beginner from arriving at the erroneous conclusion that these visible lamellar stripes and their sutures are in a sagittal direction of one another.

They in reality are concentrically situated.

Fig. 100b. The laminations of the human lens.

These surfaces have been exposed by the slitlamp. The anterior surface of the senile nucleus was first described by *Gullstrand*[1])[128]). We have observed and described additional surfaces[49])[27])[53]). When observing the laminations of the lens with the nitrogen or arclight, I would recommend the observer to reduce the width of the slit to 0,5 mm, or less.

As long as twenty years ago certain observers *L. Müller, Demichcri, Tscherning, Berlin, A. v. Szily*[50]), drew attention to faint pictures seen in addition to the *Purkinje-Sanson* images, near the anterior and posterior lens images of the latter, and correctly interpreted them as pertaining to the nuclear surfaces. *Hess*[51]) investigated them carefully and identified them as being physiologic (Kernbildchen). That these nuclear reflections give rise to incorrect conclusions regarding the number, form, location and luminosity of laminations of the lens, has been shown by the slitlamp[2])[27]). By shifting the illumination one can not only, so to say, "palpate" the shape of the *lens surfaces*, but laminations of the interior of the lens are exposed, the existence of which had not been known. Not two surfaces, as has been supposed, but a much larger number can give rise to the nuclear reflections (Kernbildchen).

We may divide these surfaces into two main groups:

      1. The embryonic nuclear surfaces

      2. The senile nuclear surfaces[27]).

Fig. 100 shows the individual surfaces as seen between the ages of 20 and 40 (sagittal optical lens section). Note the central interspace (see arrow), it separates the anterior from the posterior lens zone. On its anterior border I found[52]) a delicate typic cataract (congenital), in 25 % of old persons. (Anterior axial embryonal cataract see Fig. 244—259).

The central interspace of the lens is bounded, anteriorly and posteriorly, by a lamellar surface, in such a manner, that if a sagittal section be made, this central lens area would present a biscuit or coffee-bean shape (see Fig. 100b). Its form is approximately similar in all individuals. A very decided curving of the posterior central embryonal nuclear surface is shown in Fig. 271c. These two lamellar surfaces (5 and 6, Fig. 100b) show the embryonic sutures, anteriorly a vertical Y and posteriorly an inverted Y (Fig. 100a).

The posterior Y suture appears more luminous than the anterior one, therefore can be easily observed in all lenses. The fibre designing near this posterior suture is easily seen (see Fig. 125), the lower seams show a dichotomous branching (usually single), the upper vertical one as a rule shows no branching. On the anterior surface the lower vertical seam (suture) may only show a branching fibre design. At times the two Y seams are slightly diagonal in position, in these cases both the anterior and posterior ones are parallel to one another, that is, situated in the same meridian. I have noted these two lamellar surfaces in small children (5 and 6, Fig. 100b), in the newborn, as well as in the fetus.

The two lamellar surfaces (4 and 7), which we have termed the *second or central embryonic nuclear surfaces*, show in most persons a concentric surrounding zone and lamellar surface, which latter surface we have termed the *first or peripheral embryonic nuclear surface* (Fig. 100b, 127). These surfaces also show a simple suturing. Their formation corresponds to the time of just before birth, and at birth. These surfaces are next surrounded by a lens zone, which shows on its surfaces a suture designing composed of many branches. These surfaces are better visible in the periphery, than in the axial areas. They are also farthest from the lens surface at the periphery.

These surfaces are the *senile nuclear surfaces* (3 and 8, Fig. 100b). I have found them as early, and sometimes earlier than the tenth year of life. With increasing age they are more luminous, as a rule. In age they show a characteristic image in relief (Fig. 133—139).

Between the *senile nuclear surfaces* and the surfaces of the lens proper, there are found two other surfaces which I have provisionally named the *"surfaces of separation"*. The anterior surface of separation is more easily observed than the posterior one (2 and 9, Fig. 100b). In a like manner, as in the case of the senile nuclear surface, its distance is greater from the external lens surface in the periphery (1 and 10, Fig. 100b) than in the axial area. Its radius of curvature is smaller than that of the lens surface. At times the senile nuclear surface and the anterior surface of separation seem to merge in the axial area. By narrowing the slit of the apparatus they may be distinctly separated.

With the senile increase in the central reflection, this surface, if separation occurs, becomes less distinctly visible, in contrast to the greater visibility of all other lamellar surfaces.

In Fig. 100b, one can observe that the corresponding anterior and posterior surfaces join at the equator of the lens. This junction is however not, as a rule, distinctly seen in the embryonic surfaces in youth. With the exception of the surface of separation we are later able to observe all other lamellar zones, with increasing age. Especially more distinct and luminous with age is the second posterior embryonic nuclear surface, with its fibre formation, which latter at times may assume a wavy curled appearance in old age. Therefore the slitlamp allows us to observe the lens with its original fetal suturing in extreme age.

To be exact, the zones are not concentrically arranged, for the more peripheral the zone the greater the radius of its axial area. This proves that the lens is flattened in the course of its development.

(Ontogenetically, as well as phylogenetically, this flattening of the lens can be demonstrated in mammalia). *Rabl*,[54]) has shown this in another manner.

These lamellar zones allow of a topographic localization within the lens substance. They are therefore of importance in the further investigation of the lens by means of the slitlamp. If it be further possible to judge the age of these lens zones, that is, to estimate the time of their origin, we would therefrom have certain information, as to the age of lens clouding.

We could for instance say that an opacification situated *outside* of the embryonic nuclear zone cannot be of congenital origin, or one situated outside of the senile nuclear zone, could not have originated in infancy. (Capsular opacities, or such as may be adherent to the capsule cannot thus be utilized in estimating time of origin.) Regarding the radial extent of these various lamellar zones and their thickness exact information cannot at present be given. It is possible that we are dealing with surfaces, not sharply demarkated, with the probable exception of the senile nuclear surface. These lamellar zones are concentric areas of increasing reflection, resembling geological strata, slightly changed in physical character, from one another. With increased magnification one can observe in certain cases, that within these zones there are again further concentric layers of changed reflection, proving further separations.

I have endeavoured to ascertain anatomically the causes that lead to the development of these zones of lamellar separation, and have used for this purpose lenses, in situ, of a six month fetus, a child of 2 and 5 years, as well as a man of 29 and one of 30 years of age, and examined them according to the method of *Rabl*[54]).

Rabl[54]) had noted that the cross-section of the lens fibres is quite varied in larger mammals, more especially in primates. The investigation of these fresh lenses treated according to the method of *Rabl*, has shown that this variation does *not apply to the arrangement of fibres within the individual concentric zones. The fibres of define concentric zones show a corresponding or almost similar type of arrangement on cross-section.*

For example, the fibres directly under the epithelium, that is under the lens capsule, always show a flattened hexagonal form. This applies to the lens of the fetus, of the newborn as well as to the older ones. In sections parallel to the equator, one notes in the fetus, in children as well as in adults, a uniform zone of 15 or more fibres, on cross section. This is followed by a zone of somewhat thicker fibres, which latter zone merges with one of increased irregularity of fibres (polygonal zone). Here we find fibre cross sections many times larger, than those of the subcapsular zone, and showing varied forms.

(See microphotograph Fig. 101 = Lens of a 6 month human fetus

                Fig. 102 = Lens of a 30 year old man under increased magnification.)

Following this third zone is one of uniformly thin fibres, then one of somewhat thicker fibres, the outlines of which are less distinct. The latter merges with an extensive amorphous zone, in which the fibre outlines are difficult of recognition. In the fetus this zone is composed of large polygonal fibres. These various layers cannot be artefacts, as they can be identified by the *various methods* of fixing and hardening, nor are they found in the lenses of certain lower mammals, while constant in primates, especially in man.

By further investigations, especially by aggregate measurements, we must establish whether these zones give any anatomical basis for the lamellar separations. Not alone anatomically, but also clinically I have proven the presence of this zonular arrangement, even in the fetus.

The acceptance of an anatomic basis for the lamellar separation is supported by the uniformity of the lamellar zones of separation, as well as of the cross section zones of the lens fibre areas. As all fibres are situated originally in the subcapsular zone and show a normal regular cross-section, we reason that the fibres, after moving centralward, show in similar periods of time, correspondingly similar changes, which must then by their uniformity be characteristic of fibres of the same age. Valuable material for the study of the questions may be obtained from lower species of mammalia. I have found the optical lamination zones of the lens well shown in dogs, cats, calves, pigs and rabbits.

Figs. 103 to 111. The suture arrangement of the normal human lens and the lens fibre designing.

The arrangement of the *sutures* in the *normal lens* has not been extensively studied, and quite contradictory opinions were held regarding its shape and types of design.

The slitlamp has given us definite information, and the suture arrangement can be observed in the living eye. They are not alone of value in the determination of the lamellar separations, pathologically they also present a "locus minoris resistentiae". The sutures are intimately related to the cross section of the lens fibres, as the form of the latter is a creation of the system or arrangement of lens suturing[46]). A *correct* understanding necessitates a study of the *development* of the sutures and their arrangement.

J. Arnold[56]) has apparently proven that in the embryo of cattle the seams or sutures are arranged in the form of a three spoked star, the anterior a reversed Y and the posterior presenting a vertical Y. (Regarding the development of the sutures in pigs compare *Rabl*[54]). This observation was applied by later authors as being characteristic of human lenses. We also find in textbooks, in which the suturing system is described, the statement that in the newborn the anterior Y seam is inverted, and on the posterior surface it shows a vertical Y. The reverse is the case. Not alone in the fetus and the newborn (compare the microphotographs Figs. 103 to 111), but also in post-embryonic lenses, I have demonstrated the sutures with nitrate of silver, as being a vertical Y anteriorly, and an inverted Y posteriorly. With the slitlamp we may observe these Y (Figs. 127 and 128) shaped stars even in extreme age (Figs. 127 and 128).

The embryonic seam, as well as all other sutures, are optical cross sections of a *suture surface*, which latter is situated perpendicular to the equatorial plane. These suture surfaces do not reflect to an equal degree. If one or more of the cross sections of these suture surfaces show a maximum reflection, they will create a suture designing or pattern. (This is also the case in other mammals. More distinct than in human lenses is the suture arrangement in cattle, pigs, cats, dogs and rabbits.) Areas that are less luminous by letting the reflection wander may be observed as presenting a definite designing. The posterior embryonal suturing and its fibre formation is the most distinct of the two. Their characters may differ, the less luminous anterior suturing may be broader and, as a rule, shows no branching, or only on its lower two arms (Figs. 104 to 106). The posterior suturing may show a forking at the end of the spokes, the upper vertical one seldomly shows branching. At times the Y is diagonal, in the anterior as well as in the posterior suture.

Fig. 100—109. Tafel 11.

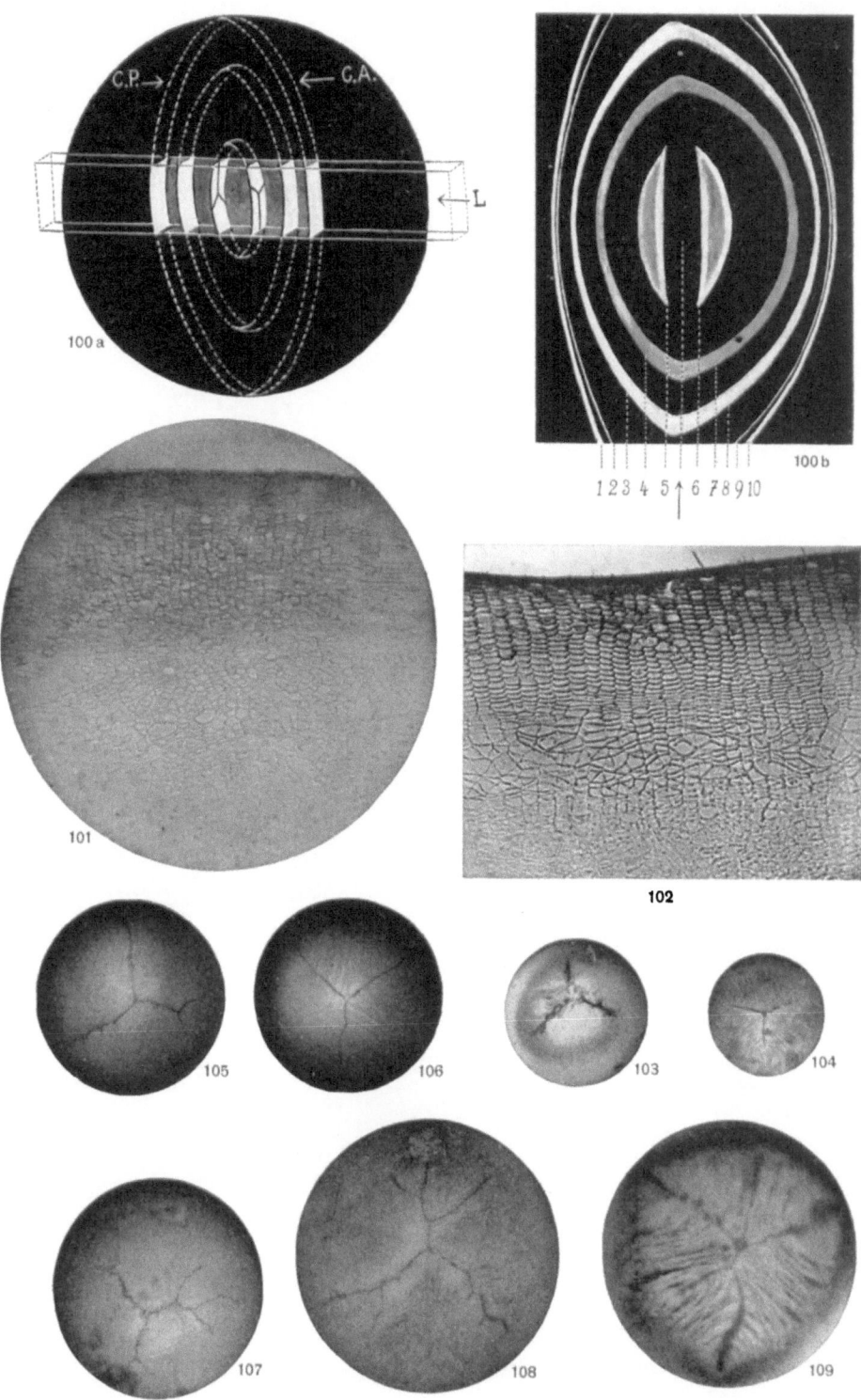

In the newborn the posterior central junction of the three spokes (sutures), corresponds fairly well with the location of the posterior lens pole.

During life it remains situated at the posterior axis, which fact is of importance for the purpose of localizing this area, as well as other areas within the lens substance. Regarding the direction of the seams (sutures), and the branching, we may add: The branching is usually dichotomous, both branches resembling *gothic curves meeting at a point* (Figs. 92b, 123, 130 etc.). Ontogenetically the branches at their beginning are slightly kinked. (Compare Fig. 106, 116 etc.) This kinking creates the antler formation as illustrated in Figs. 113a, c, 115a, b, c.

Microphotographs and sketches of the sutures in lenses at various ages:

Fig. 103. Human fetus 4—5 months, posterior suturing with nitrate of silver impregnation. Owing to slight maceration of the lens the suture seems broad and less distinct. Branching not visible for this reason.

Fig. 104. Anterior sutures of a human fetus 4—5 months. Distinct seams, no branches, lens fibre formation distinct.

Fig. 105. 7 months—posterior suturing, right lower seam branched singly, the left lower one shows a kinking. (The branching according to our observations is inaugurated by a similar kinking, this latter involving the lens fibres in apposition.)

Fig. 106. 6 months—anterior suturing, no branching, the lower seam (suture) shows a kinking in anticipation of a branching. Lens fibres distinct.

Fig. 107. 7—8 months. Exceptionally frequent branching of the posterior suture. The left lower seam shows a double branching. (In this as well as in other illustrations the axial seam junction is not exactly central, owing to the slightly diagonal position of the preparation.)

Fig. 108. Posterior suturing in neonatus. Lower seam shows branching only, as is characteristic of this surface in the newborn. Owing to absorption of fluid, an apparent seam is seen upward (artefact due to maceration), in the picture. The fibre direction also shows that this is not a seam.

Figs. 110 and 111. Illustration of the suturing arrangement and the posterior tunica vasculosa in fetuses, 4 and 6 months. These eyes had been kept in dilute alcohol, at the time of section, the vitreous was watery and flocculent. The lenses were placed in a 1—1000 solution of nitrate of silver for 15 minutes. Note the entrance of the arteria hyaloidea external to the central junction of the seams (sutures).

Figs. 112 to 121. Observation of the suturing (lenses of recently deceased) with the slitlamp, low magnification.

Impregnation of the seams by a 1—1000 solution of nitrate of silver. It was not always possible to so accurately arrange the lenses, in situ, that the suturing would show in its regular arrangement. The relation of length of sutures and equatorial radius was only ascertained in certain cases. The radius of the sutured area was approximately three quarter of the radius to the equator. The seams (sutures) extend farther into the periphery in age. In the latter there is often a change in the relation of the anterior to the posterior suturing, in this respect.

Fig. 112a—f. Superficial sutures, 2 fetuses, 27th—28th week, *a, c, e* posterior, *b, d, f* anterior.

Fig. 113a—d. Superficial suturing of neonatus, *a, c* posterior, *b, d* anterior.

Fig. 114a—f. Superficial suturing in two newborn, 2 and 3 days old, *a, c, e* posterior, *b, d, f* anterior.

Fig. 115a—c. Superficial suturing boy age 11 days, *a, b* posterior, *c* anterior.

Fig. 116a—f. Superficial suturing, of 3 children 7—8 weeks of age, *a, d* posterior, *b, c, e, f* anterior. Notice in Fig. 116f the branching is limited to a single seam (probably the lower one). The lower seam shows, as has been stated, the first branching in the anterior suture arrangement. (Fig. 106, 109, 112b, 113d, 114d, 114f, 116b, 116c, 116f.) It gives a basis for the assertion that this branching gives origin to the *vertical suture ridge* of the *anterior senile nuclear relief image* (see Fig. 137, 139).

Fig. 117a—g. Superficial suturing of two children age 9 and 10 months. The branching is well advanced, *a, c, e* posterior, *b, d, f, g* anterior.

Fig. 118a—d. Superficial suturing in a child of $2^3/_4$ years *(a and b)* and one of 3 years *(c and d)*, *a, c* posterior, *b, d* anterior.

Fig. 119a—d. Superficial suturing in a child $3^3/_4$ years old *(a)*, one of 5 years *(b, c)* and one of 7 years *(d)*, *a* posterior, *b, c, d* anterior.

Fig. 120a and b. *Anterior superficial suturing* at 16 years *(a)* at 60 years *(b)*.

Fig. 121a and b. Superficial suturing at 78 years, *a* anterior, *b* posterior.

Fig. 122. Anterior nuclear superficial suturing. Mrs. T. age 63.

Sketched from life—microscope and slitlamp.

Fig. 123. Microphotograph of the anterior suturing arrangement at 50 years.

After impregnation of the sutures of the fresh lens, in situ, with a 1—1000 solution of nitrate of silver.

Fig. 124. The same as Fig. 123, anterior seam at 60 years.

The number of main branches varies from 6 to 10 in adults. The smaller branches vary from 10 to 15 in number.

(The lenses of recently deceased were obtained from the department of anatomy and pathology of the University of Basle—Director Prof. Hedinger.)

Figs. 125 and 126. Anterior and posterior embryonic suturing, at 30 years of age.

Oc. 2, Obj. a2. The lens fibres radiating from the posterior suture are stretched. Dichotomous branching of both lower sutures. Note the straight pointed form of the anterior, in contrast to the kinking in the posterior radiations. The posterior suturing as a rule shows more luminously, and is therefore easier of observation. In youth it shows light on darker background, in age, darker on a more luminous background. A great number of measurements gave for the posterior suture (to the first branching) an apparent length of $1^1/_4$—$1^1/_2$ mm, and for the anterior lower sutures one of $1^1/_2$—$1^3/_4$ mm. The branches of the posterior sutures I have found in some cases to be $^1/_2$—$^3/_4$ mm in lenght. If one accepts that the volume of the lens nucleus remains practically stationary in the first year of life, within the given measurements, one must agree that the embryonic nucleus corresponds in size to the lens of the *last fetal month*, that is the *time of the beginning of retrograde metamorphosis of the tunica vasculosa lentis.* The end of embryonal life therefore remains demarkated through life in the lens by this lamellar zone. It in a certain sense represents an annular ring of growth ("Jahresring").

Fig. 110—118a.

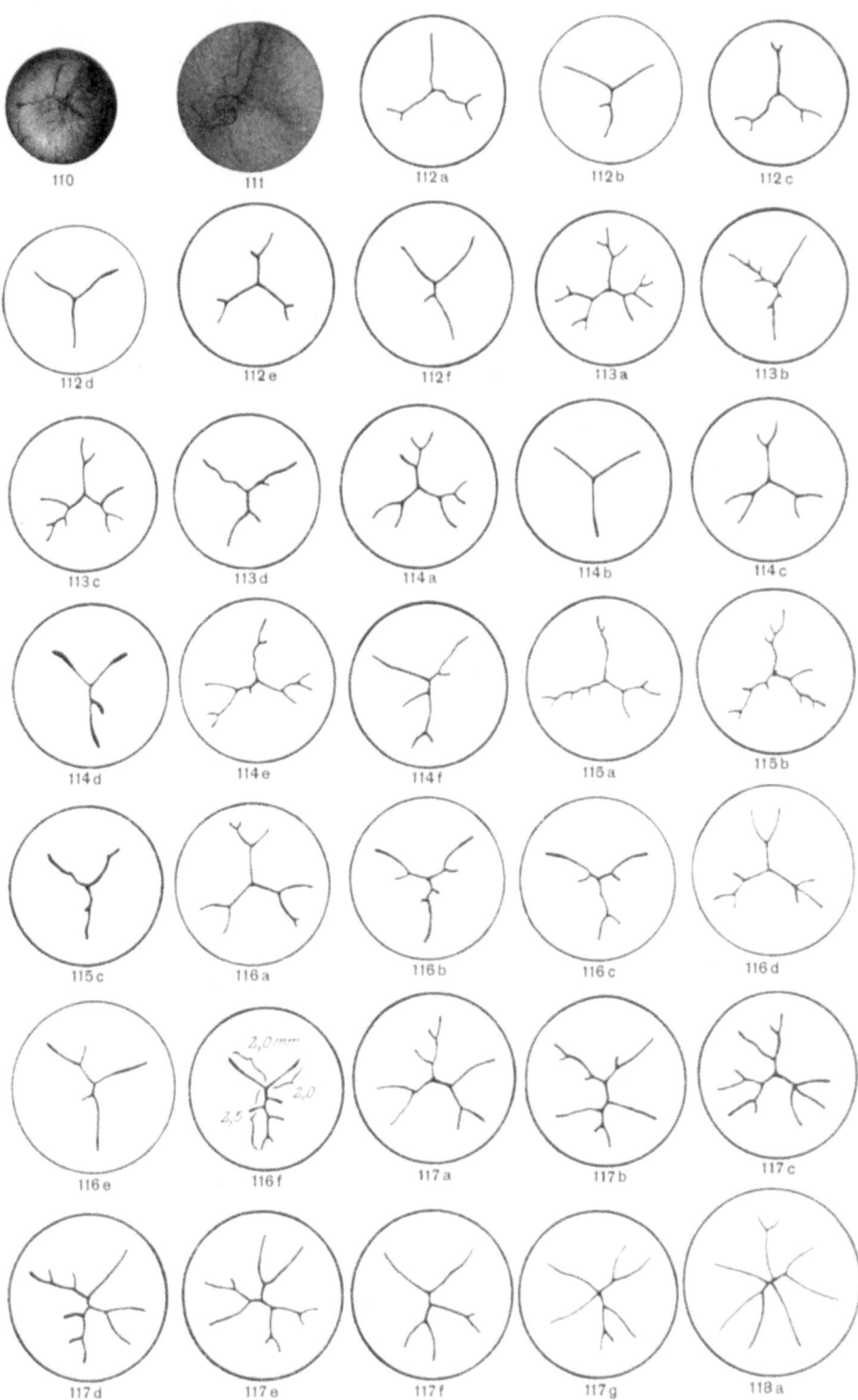

Verlag von Julius Springer, Berlin.

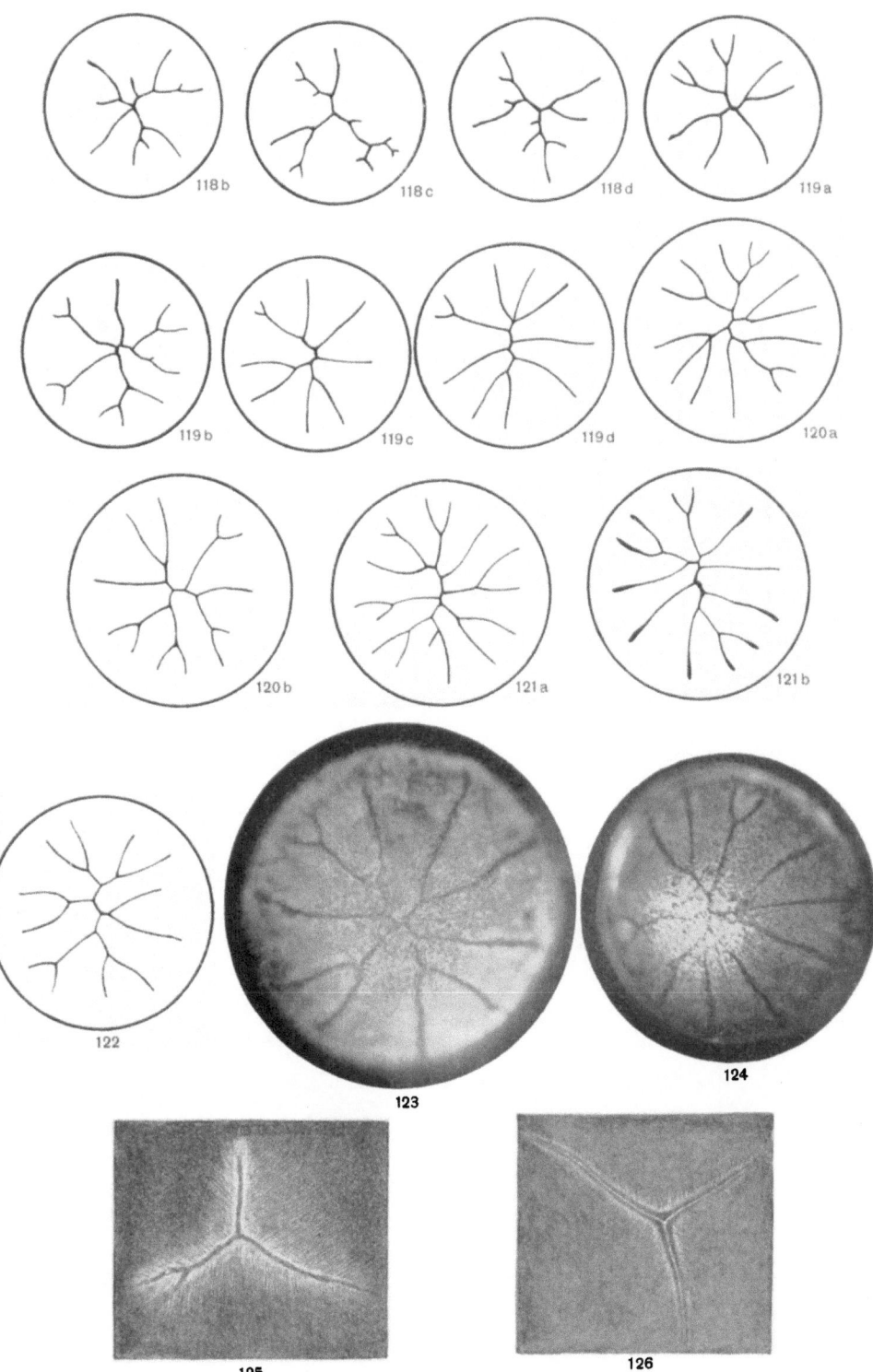

118b

118c

118d

119a

119b

119c

119d

120a

120b

121a

121b

122

123

124

125

126

Verlag von Julius Springer, Berlin.

Figs. 127. Anterior and posterior embryonic suturing in a child.

Oc. 2, Obj. a2. Note the luminosity of the sutures on the dark background. The embryonic suture in contrast to this, in age as a rule shows dark on a luminous background, the opposite is true of the sutures of the cataractous nucleus (see Fig. 229).

Fig. 92. Anterior suturing of the cortical surface, age 20.

The anterior lens graining in focus (see page 20), shows the suturing of the anterior lens surface. Its visibility varies in individuals, and is more easily observed in youth. It is especially distinctly seen in cases of traumatic cataract. The suturing on the posterior lens surface, owing to the reduced size of the reflecting image, is only visible in small areas, and as a rule only at the periphery (see Fig. 98).

Figs. 128 to 132. The width of fibres, its dependence on the arrangement of sutures.[46])

The schematic sketch Fig. 128 illustrates this dependence. In the triangle, the base is represented by "x", the two sides sutures by "a" and "b". The fibres inserted at the sides "a" and "b", being situated in one plane must all pass the line "x". If the length of $a = 2x$, then $a + b = 4x$. Therefore at the origin of the fibres in the line "a" and "b", they must be four times as broad as when the same fibres pass in the plane "x". These differences increase in the same proportion as "a+b" is greater than "x". The more numerous and longer the sutures in relation to the size of line "x" (if, for example, we erect the vertical added suture "c" at "x"), the greater the fibre diameter at the sutures in proportion to their diameter when crossing the plane at "x". In the last named case the relation would be as 8 to 1.

This is the case where the surface between sutures is a flat plane. The fact that the lens surface is curved, does not materially change conditions. The microphotograph Fig. 130 which presents the anterior superficial suturing and the fibre limits made distinct by nitrate of silver, in an elderly woman, shows conclusively that the length of the suture is many times greater than the surface area which they must occupy in the periphery, therefore the fibres must greatly increase in their width, at their junction with the sutures, as shown schematically in Fig. 129.

Fig. 131 and 132. *Microphotographs of normal fresh adult lenses illustrating the widening of the fibre ends.* This condition has in the past been overlooked (it has been stated however that the fibres at their ends show a bulbar thickening), owing to the fact that only cross-sections were being examined. We *cannot* fully explain matters in this way, as I have proved by a series of sections of human lenses, treated after the method of *Rabl.*

Fig. 131 shows the fibre ends as they approach a surface suture, therefore slightly curved, in a fresh lens of an adult under high magnification.

Fig. 132 shows the superficial fibre surface treated with nitrate of silver. I obtained very instructive pictures by a mild impregnation with nitrate of silver, then after hardening a short time in alcohol, I scraped off the cortex. The capsule then showed the impressions of the fibre ends in a very distinct manner. As the equatorial line of the lens is greater than the equatorial line, which can be imagined at the ends of the sutures, it follows that the fibre width is greater at the equator than at the fibre ends (a certain compensation is due to the fact that the lens curvature is increased toward the equator). In the lens preparation of a boy aged 5 years (prepared after *Rabl*), I found the relation of the width of the fibre at the suture ends to that at the equator as about 3 to 4 (correctly $13\,^1/_2$ to 18).

It cannot be a coincidence that the greater development of the sutures phylogenetically as well as ontogenetically corresponds to an *increased flattening of the lens* in the dorso-ventral direction.

As the fibre ends of the axial lens zone do not belong to the equator, there would then result a reduction in the antero-posterior lens diameter. It is a fact that a meridional section of the adult lens shows the fibres thicker at the equator in a radial direction, than axially. Diagonal fibre sections also show this. (Others have claimed that the *fibre nuclei* are responsible for the flattening of the lens. They cannot play an important part as they are equally present in lenses presenting *no flattening.*)

In accord with the above, one must presume that the flattening of the lens corresponds to the development of the suturing, an observation proven by facts, as the embryonic lens is almost spherical. The longer and more numerous the sutures, the greater the difference between the fibre width of the suture compared to that of the equatorial area, and as a result the greater the degree of dorso-ventral flattening. (The same is a fact phylogenetically and ontogenetically for the lenses of mammalia.)

Figs. 133 to 139. The relief image of the senile nucleus.

In the year 1913 we succeeded in proving the presence of a relief image of the anterior axial senile nuclear surface in all persons over 40 years of age.[46] [2] [57] [58] [27] One may see well developed relief images, if well illuminated, by the use of the *Hartnack* loupe.

With the slitlamp and corneal microscope (Oc. 2, Obj. a3) we see relief images as shown by Fig. 133 to 139. (Photographs of plastic models made by Dr. *U. Lüssi*, assistant ophthalmologist at the clinic.)

In other cases the relief picture is less distinct, one can see the seams as dark lines on a flat surface, or only the suture ridges. The anterior relief image consists of ridges, banks, and knobs arranged in certain relation to the sutures and fibres. The latter show as ridges (for instance Fig. 135). At times the fibre designing is visible (Figs. 135, 136, 139). A vertical suture ridge in the axial area is especially typic of the anterior relief image (Figs. 137 and 139, also Text to Fig. 116). The posterior relief image (Fig. 138) shows in a similar manner as the anterior, if the latter were viewed from the rear. Prominences appear as depressions etc.

This can be proved by variations in the direction of light, by the behaviour of reflections and shadows. Axially the relief image is more distinct, and the branching better visible than in the periphery.

Pathologic changes show as round, sharply circumscribed mounds (Figs. 133, 137). They are comparable to round-headed shoe-nails, and are due to vacuole formation, directly under the nuclear surface. We see them in the anterior as well as in the posterior relief image[58]. By *transillumination* the relief images are invisible. The *visual acuity* is not noticeably affected by them. *Lenses showing a very distinct relief image have been observed in eyes having a visual acuity of 1,5 ($^{15}/_{10}$).* We could not say that they bear any direct relation to cataract formation. Only in a few cases have we found transitions in the direction of lamellar separation. One can study the anterior relief image through a narrow pupil. The pupillary border and the common physiologic deposits on the anterior capsule (see Fig. 141 etc. remnants of the pupillary membrane), or opacities of the cortex, which may possibly be present, throw their shadows onto the relief image and show parallactic displacement with one another, when seen by slitlamp illumination.

Fig. **127—133**.   Tafel **14**.

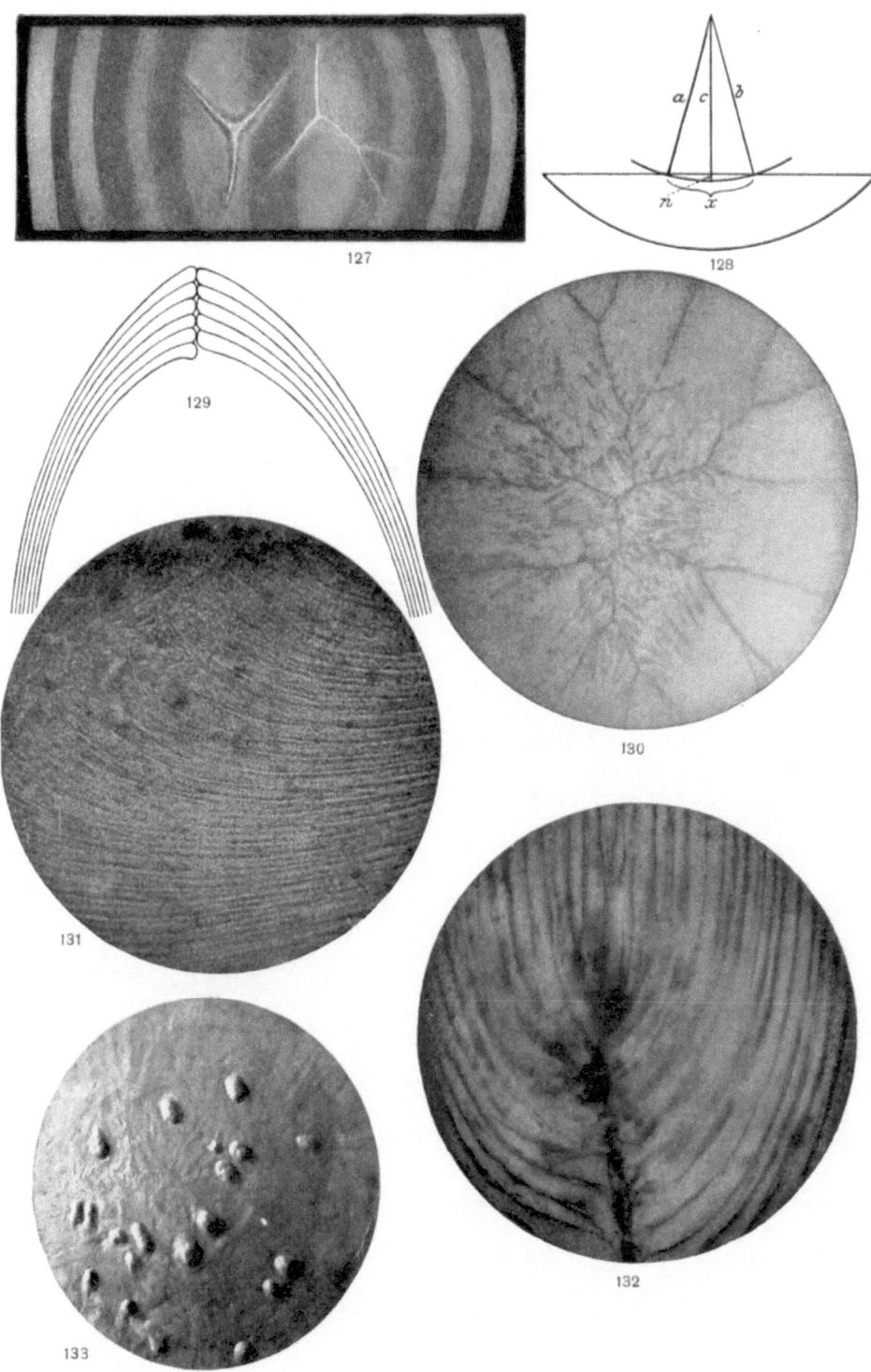

127

128

129

130

131

132

133

The relief image is definitely differentiated from lens opacities as follows:

1. The relief image is invisible by transillumination.

2. Confusion is impossible because of its characteristic form, when once seen.

3. By its axial location, and by its being on the *senile nuclear surface*, anterior as well as posterior.

4. By the involvement of the sutures in ridge formation.

5. By the continuity of the image. The relief image is continuous in all directions, flattening and gradually disappearing toward the periphery.

At times I found a relief image in the *anterior embryonic nuclear surface* (see Fig. 139).[27] [58])

Both types, that of the senile nucleus, and of the embryonic nuclear surface may be present in one and the same lens. (Regarding the frequency of occurrence of the various forms see [58].)

Fig. 133. Male age 84. R.V. = $^{6-7}/_{200}$ nuclear cataract, pupil dilated, light temporal. No ridges on the anterior nuclear surface. In the centre several circumscribed mounds on a flat base. They are round and oval, their surfaces and the surroundings smooth, no confluence. They differ from the usual knobs found on the relief image by their shape and increased size.

The left eye of the patient shows the same condition. Extraction of the lens in this eye brought forth a large nucleus; clear cortex was left behind. Immediate examination of the nucleus with the corneal microscope failed to show the mounds seen in life. (Double nuclear cataract, peripherally a few small lens opacities. L. V. = $^{6-7}/_{200}$.)

Fig. 134. Male age 82 (LV $^{3}/_{200}$ nuclear cataract). Pupil dilated, light temporal. The whole of the anterior nuclear surface shows mounds and knobs on a bright, slightly dull luminous base. They are united in groups, here and there confluent in contrast to Fig. 133. No suture ridges. The knobs continue to the pupillary border. (The fine striping occurred in creating the plastic image, it was not present on the anterior nuclear surface.) (R aphakia-central chorio-retinitis V = $^{9}/_{200}$ with correction. L coloboma of iris up, nuclear sclerosis, subcapsular cortical opacification.)

Fig. 135. Female age 72 (RV = $^{1-2}/_{200}$ amotio retinae). Pupil dilated, light nasal. Typic relief image of the anterior nuclear surface, some of the details visible with *Hartnack* loupe (knobs, lens fibres, ridges). The central area of knoblike elevations, no vertical ridge, shows the sutures in a radial direction, extending to the pupillary border. A lower temporal suture ridge bifurcates in the characteristic way. Between the ridges of the sutures are the fibre fields forming a uniform striped relief image.

The typically arranged fibre connections at the sutures often cover the ridges (right and upward). The grooves between the fibres in certain areas are quite deep (right lower field). The fibre ridges are irregularly thick in places (right field). Note the irregular grooving and swelling in the axial area, showing knobs and in places the fibre formation. The suture ridges are accompanied by grooves, in which the fibre formation shows. (Double nuclear cataract, cortex clear, except spokes. LV = $^{4}/_{200}$.)

Fig. 136. Male age 58 (LV = $^{6}/_{24}$ incipient nuclear sclerosis). Pupil dilated, light temporal. Similar relief image as in Fig. 135 with less distinct ridge and knob development. Below the area showing knobs, there is an irregular groove between the fibres. (R spoke opacification of cortex, subcapsular vacuolar surface, L same-increased nuclear sclerosis.)

Fig. 137. Male age 72 (LV = $^6/_{36}$ corr. to $^6/_6$). Pupil dilated, light nasal. The whole pupillary area shows a uniform extensive field of variously formed knobs. The axial vertical ridge shows a wide central separation. Its upper and lower edges each give off three fan-like sutures. They are covered here and there by fine knobs and grooves. Nasal and upward there is a large mound on the uneven surface. It differs only in size from the one in case 133 and 134. With a certain illumination this mound shows a vacuole-like reflex. The posterior relief image of this nucleus is relatively less well developed, it shows three vacuoles and they give rise to a similar mound-formation as shown in Fig. 133 and 134.

(R herpes zoster ophthalmicus and quiescent corneal opacification. L relief image of anterior and posterior nuclear surfaces (Fig. 223), lamellar separation in the deep cortex. Focal illumination with the ophthalmoscope shows the lens clear. RV = $^5/_{200}$. This case was under observation for one year. In this time there was no change in the relief image, however the lamellar separation increased.)

Fig. 138. Male age 64 (LV = $^6/_{24}$ with corr. $^6/_6$). Pupil dilated, light temporal. Relief image of *posterior* nuclear surface. Deep axial vertical groove. The side of the groove nearest the source of light is in the shadow, the side farthest from the light is luminous, as is characteristic of the appearance of groove formation. Similar to suture ridges on the anterior relief image, there are radial deep grooves at both ends of the vertical suture. They are quite short and present the same phenomena of light and shadow as the vertical suture itself. The longest grooved suture is situated downward and temporal. It becomes indistinct where it divides. The corresponding nasal and downward grooved suture is indistinct because of a spoke in the cortex posterior to it. The whole zone of groove formation lies in a field pitted in varied sizes and depths. These pits correspond to the knobs of the anterior relief image and show a similar behaviour toward the light as the grooves.

The knobs as well as the pits are most distinctly seen axially, and decrease in number and depth toward the periphery. The whole picture shows parallactic displacement in relation to the deeper, more luminous lambdoid suture of the embryonic nucleus. (R keratitis profunda, following herpes zoster ophthalmicus. Tumour of choroid? bilateral peripheral cortical opacification. Vis = $^6/_{18}$ with correction.)

Fig. 139. Spinster age 43 (RV = $^6/_6$ Hm 1,5). Pupil dilated, light temporal. Relief image of the anterior embryonic nuclear surface. The Y shaped sutures are prominent as ridges. The vertical seam near the centre shows a nasal kinking. Near and at the suture ridges there are wavy knobs, which latter are lost into the fibres toward the periphery. In the area of knobs and on the ridges the fibre designing is somewhat curled.

Fig. 140a and b. Increase in the internal diffuse reflection of the lens.

To determine an increase in the internal lens reflection is somewhat difficult with slitlamp-microscopy. It is relatively easy to determine a localized increase in an individual lens, for instance, an increased internal reflection of the nucleus, in comparison to the external lens substance.

Extremely difficult however is a comparison between two individual lenses, as the mental impression of luminosity given by one lens is lost by the time the second lens is brought into focus.

The conditions may also vary. The luminosity of a lamellar area is dependent on whether or not the line of observation corresponds to the direction of greatest reflection of the illuminated surface. Without doubt the inner reflection of the lens

Fig. **134—139.** Tafel **15.**

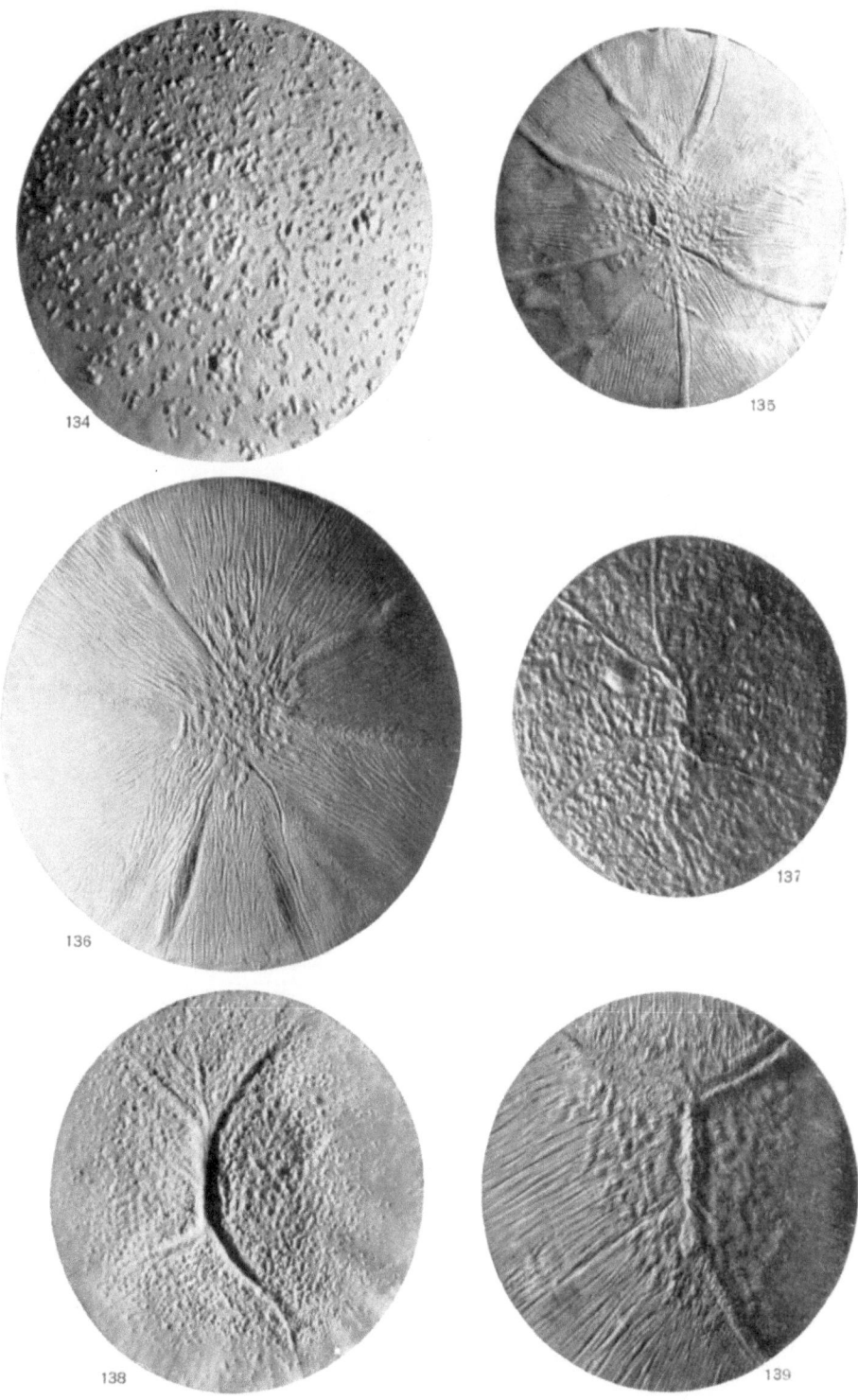

increases with age. The various lamellar surfaces increase in their luminosity, so that we may compare lenses of dissimilar ages (Fig. 140). The lens at 6 years "*a*", the lens at 80 years "*b*". The contrast is sufficient to remove all doubt, at least if the light is not too intense.

To compare lenses it would be necessary to have pictures taken under similar conditions of illumination showing the various degrees of average luminosity at given ages. An increase in the diffuse internal reflection is said to be the cause of socalled night-blindness, according to *Koeppe*[59]). We have had no occasion to investigate this, but the assertion seems contrary to fact, as there is no hemeralopia, though the internal reflection *increases decidedly* in age. An increase in the internal lens reflection may be due to cataract, without the presence of circumscribed or dust-like opacities. This is the case in nuclear sclerosis.

Fig. 140a and b. The increase in the yellow colouration of the lens in age.

In contrast to the lenses of calves, the lens of the human newborn is slightly yellow. In addition to ultraviolet, violet and blue are also absorbed and they give rise to fluorescence (compare Vogt[10]).

In old age the absorption of colours is increased and it includes blue and green, so that the yellow colouration finally reaches an orange or reddish-yellow.

This may give origin to blue blindness (*Liebreich*[109]), *Hess*[110]) and others), and a relative yellow blindness (Vogt[111]).

Fig. 140a shows the colouration in an individual 6 years of age, and Fig. 140b in one 80 years old.

The yellow colour increases from the cortex toward the nuclear centre very likely in relation to the greater density of the lens substance. If the bundle of light be directed axially through the lens, the *posterior cortex* will appear a more pronounced yellow, owing to the fact that it is seen through the layers anterior to it, during which process the light is (doubly?) filtrated through these layers (Fig. 140a and b). We may thus determine the degree of yellow colouration of the lens by means of the slit-lamp. The yellow colouration as is well known brings about the blue vision in aphakia.

Figs. 141—146. Remnants of the anterior tunica vasculosa lentis (pupillary membrane).

This membrane not alone receives its vascular supply from the hyaloid artery, which latter sends parallel branches around the lens equator to the anterior lens surface, there forming a plexus under the iris, but also is supplied by (ciliary) vessel branches from the iris.

With the slitlamp we find remnants of these vessels, postembryonic on the iris stroma, as well as on the anterior lens capsule. The latter are most conspicuous.

Fig. 141. Star shaped remnants of the pupillary membrane.

Oc. 5, Obj. a3. There are starshaped brown to brownish and yellowish white deposits, often arranged in rows or chains (comp. *Brückner*[60]), *Stähli, Kraupa*[146]), *Koeppe* and others), at times singly, in groups, or in flat piles. The individual little stars usually have a diameter of 20—60 microns. Fig. 141 shows the stars under high magnification in J. H. age 69. Note the thin radiations.

The embryonic pupillary membrane contains no pigment. As we often find pigment in various quantities in pupillary membrane remnants, we must conclude. that this pigment is later on developed in these remnants (compare *Brückner*[60]).

We must emphasize that the posterior capsular membrane *never* presents pigment, while the pupillary membrane usually shows it. This probably is an expression of the fact that the posterior vascular supply is derived from the arteria centralis, the anterior of the latter, as well as of the ciliary vessels.

The starlike shape of these remnants (Fig. 141) is not necessarily characteristic of their embryonic origin. According to our observations, iris pigment and exudates on the lens capsule, following injury and inflammation, after many months and years, may show a like appearance (compare Fig. 189, 190).

On the other hand congenital pigment may present an amorphous appearance, (For instance in Fig. 145.) Minute star and thread shaped remnants of the pupillary membrane are shown by the slitlamp in all eyes.

Fig. 142. Thread to spiderweb-like remnants of the pupillary membrane.

These structures are always connected with the iris stroma and may be attached to the anterior lens capsule, or may extend across the latter without attachment. At times some of these pupillary threads give off a branch, which latter may be connected with the capsule.

More rarely there is to be seen a delicate meshwork. I noted one of a very dense membranous structure, it was circular in shape and intimately connected with capsule and iris stroma. At times dense whitish deposits and pigment debris are attached to these threadlike fibres. These fibres cross the pupil in the manner of telegraph wires. Regarding polymorphous structures of the iris stroma, of this nature compare chapter describing the iris.

Rarely do we find these threadlike remnants adherent to the pupillary border. They may be traced to the membrana capsulo pupillaris (compare Brückner[60]).

The same illustration shows *membranous remnants of the pupillary membrane.*

Their form shows the direction of the embryonic vessels. They are deposits on the anterior capsule which simulate the *form of the vascular loops of the embryonic vascular membrane.*

A comparison with Fig. 143 (the microphotographs were obtained by transillumination following injection by Berlin-blue, which shows these vascular loops in a 5 month fetus and Fig. 144 the same in a 4 day old kitten, illustrates an exact likeness in form. Fig. 142 shows the grayish white area to be a delicate, membranous, uniformly thin and translucently dull deposit. At its margins this deposit is free and slightly everted in certain places (to the right side in Fig. 142). In the central area there are brown pigment remnants on the anterior capsule. The case shows the left, heretofore healthy eye, corresponding to number 137, which is the right eye, under a magnification of 24 times. There is slight pupillary dilatation.

Fig. 145. Amorphous (not star shaped) congenital pigment, on the anterior capsule. Mrs. L., age 43, no eye disease.

Oc. 2, Obj. a3, right eye. Aside from a few whitish remnants, suggesting starlike shapes, there are two large irregular clumps, covered by a grayish mass.

Fig. 146. Solid remnants of the pupillary membrane.

These, in a manner similar to an exudate, almost give rise to an occlusio pupillae. Characteristic however is their connection with the iris stroma. Fig. 146 especially shows numerous **membranous remnants**, which suggest posterior synechia,

Fig. **140—147**.   Tafel **16**.

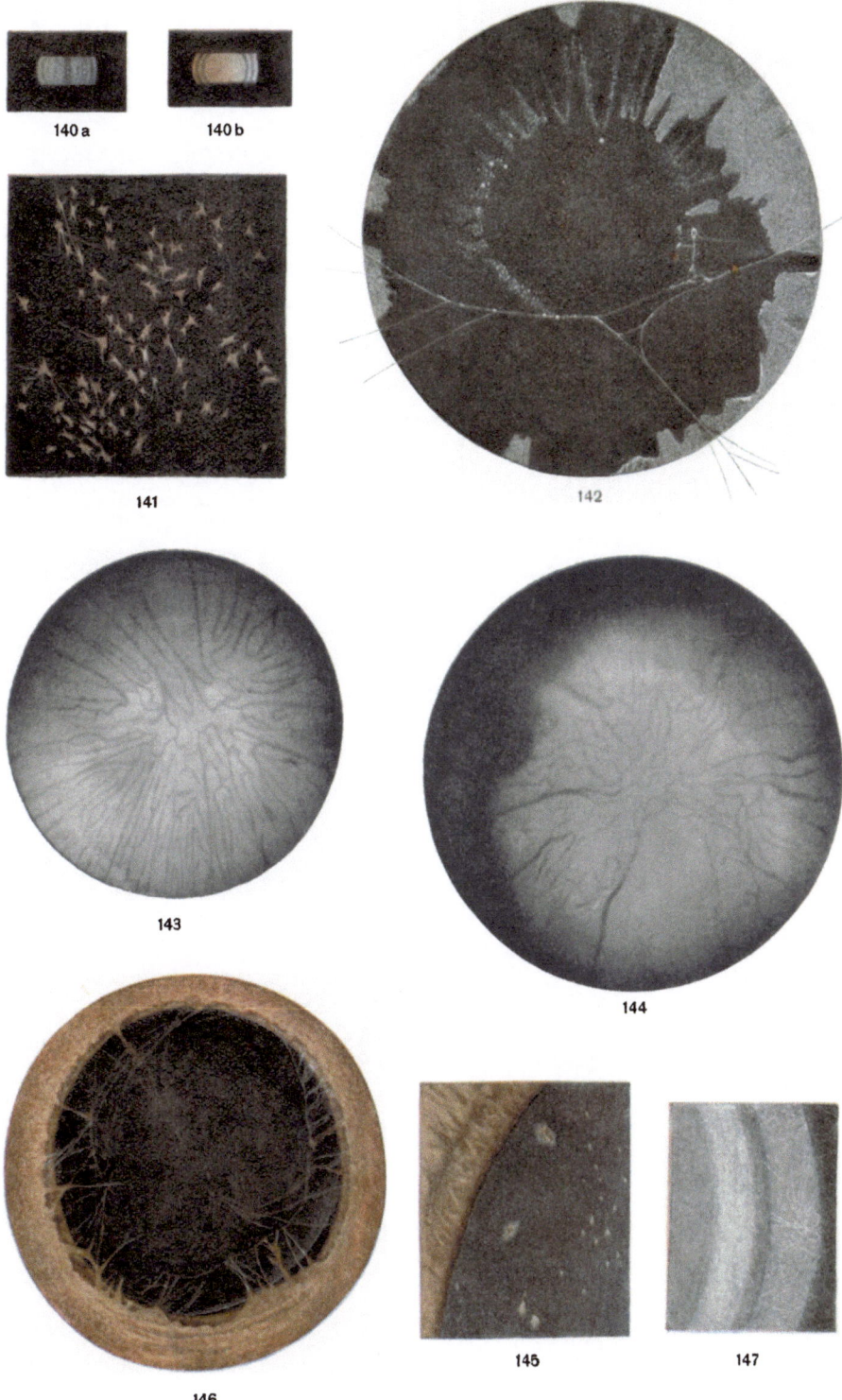

140 a   140 b

141

142

143

144

146   145   147

however they disclose their nature by their connection with the iris stroma. The lens in this case is luxated upward congenitally (Miss E. age 10, right eye, low magnification).

With these illustrations we have shown only a few of the most usual types of remnants of pupillary membranes.

Figs. 147 and 147a. Remnants of the posterior tunica vasculosa.

Before the introduction of the slitlamp these structures were unknown. We can show[49]) that they occur more or less numerously in all individuals. These observations have been verified by others *(Koeppe[51]).*

Fig. 147 shows threadlike remnants of the temporal area on the posterior capsule of an individual at 20 years of age. Their appearance makes it possible that they are remnants of vessels. They are mostly threadlike structures, having a radial and slightly curved direction. (Compare with the direction of vessels of the embryonic membrane, Fig. 148—150.)

These remnants can be observed much increased in number and with greater distinction with the micro-arc and nitrogen slitlamp, than with the Nernstlamp. In domestic animals I found these remnants better preserved than in man. Fig. 147a shows the vascular remnants of a fresh calf's eye "*N*" suture „*A*" branches of vessels.

Fig. 148—163. The physiologic remnants of the arteria hyaloidea.

The *arteria hyaloidea is not inserted* at the *posterior lens pole*, but according to our observations[62]) in the human fetus, *in a slight nasalward direction*.

We have subsequently found that *Seefelder*[63]) had given first *anatomical* proof of its nasal insertion in the human fetus.

This entrance corresponds, according to our opinion, with the medial site of the entrance of the optic nerve. An entrance of the optic nerve in the middle line would be optically impossible. A central insertion of the hyaloid artery would be a disadvantage as, having been proven by the slitlamp, there are always more or less dense remnants of this vessel present at the site of attachment, therefore this would interfere with the axial direction of the line of central vision. It is a fact we have proved in living individuals that the postembryonic hyaloid artery remnants can also be found 1—2 mm nasalward from the centre of the lens. To locate the posterior lens centre, one must find the central junction of the posterior embryonic suture (embryonic suture Figs. 151 to 164).

The hyaloid remnants can be observed in most persons, however their form and extent are subject to marked individual variation. They are seen especially distinct, as are all remnants of the tunica vasculosa, by observing with the micro-arc-slitlamp. The distribution of the vessels in the posterior tunica vasculosa is shown by microphotographs Figs. 148 to 150. I have obtained these specimens from living 5 month fetuses, injected with *Berlin-blue*. (I kindly thank Prof. Labhardt, and PD. Dr. Hüssy, Basle for the several fetuses I used in this work.)

Figs. 110 and 111 show the topographic relations between hyaloid artery and the superficial suture arrangement in the fetus. *Post embryonic* I have found the remnants to *vary greatly* in development in different individuals. Some were just visible with the slitlamp, while others were of a very marked intensity, so that they could just be seen with the ophthalmoscope. These dense remnants, visible by transillumination, are known as *cataracta polaris posterior spuria*, or as *hyaloidea persistens*. As we have stated and as can be proved by the slitlamp, these cataracta spuria are not situated at the posterior pole, but they are nasalward and slightly downward in the area of the remnants of the hyaloid.

Figs. 151—163 show various *common types* of the postembryonic physiological hyaloid remnants. They have been selected from about 100 cases. (To determine the position of the posterior embryonic suture see Fig. 151 *E*.) The posterior embryonic suture is always situated temporalward of the hyaloid artery remnant.

There are two types, easily differentiated:

1. The site of *fixed attachment* ("*a*", Figs. 151, 152).
2. The *floating hyaloid* (in the vitreous) ("*b*").

In the same individual the relative amounts of fixed attachment and floating remnants may vary in the two eyes. However as a rule paired eyes show a marked similarity. Let us first examine the *area of fixed attachment.* We find it by locating the posterior embryonic suture (the inverted Y, see illustration). With a little practice and by a slight dark adaptation we may see this suture in any individual. We now follow the suture seam which is directly nasal and slightly downward until it shows a forking division. The attachment zone of the hyaloid is very near the upper of the two branches, or somewhat behind it. The area of attachment as well as all other delicate structures of the posterior capsule are visible only *outside* of the *posterior reflecting zone.* They are covered by the latter, hence invisible. If the posterior reflecting zone is on the area of attachment, the patient must be directed to look slightly nasalward. This will shift the reflecting zone toward the nasal side. The area of attachment is characterized by grayish-white curved lines, which form a plexus, so that we may speak of an *attachment plexus* (Fig. 158—161). This structure, by virtue of its configuration, reminds of the embryonic attachment area with its crisscrossing and forking vessels. The width of the vessel remnants appears to be from 0,03 to 0,06 mm.

In places they may be ringformed or remind of scrolls. Similar curved lineal remnants may exceptionally be found at other places on the posterior capsule, in these cases they are smaller, and more sparse than at the area of attachment. Often it is clearly defined that the curved lines at the area of attachment originate from *various* vessels, corresponding to the various branches of the artery. The presence of skein and button-like thickenings (Figs. 156, 159, 163), as shown by our observations, give origin to the *cataracta spuria*, quite commonly found.* They may be on the capsule or attached to the floating hyaloid remnant. In the latter case they move about in the vitreous, their swaying can only be observed microscopically in its relation to a fixed suture. If they are situated on the hyaloid remnant at some distance from the area of attachment, their motility may be observed with the slitlamp, by certain bulbar movements. In this case they are also visible with the ophthalmoscope.

As a rule these remnants are bulky, more rarely flat deposits on the capsule. At times they form the proximal end of the hyaloid artery, so that the latter appears inserted into these remnants (Fig. 163). The universal similarity in size of these remnants is quite characteristic.

The second part of this attachment area, the before mentioned floating spiral-shaped white strand (Fig. 151 and 152 at b), is most noticeable on bulbar motion, whereby it is set actively into pendulum or whip-like gyrations. This proves that its specific weight is greater than that of the vitreous.

---

* These thickenings are probably identical with the white structures first described by *Erggelet* [102]).

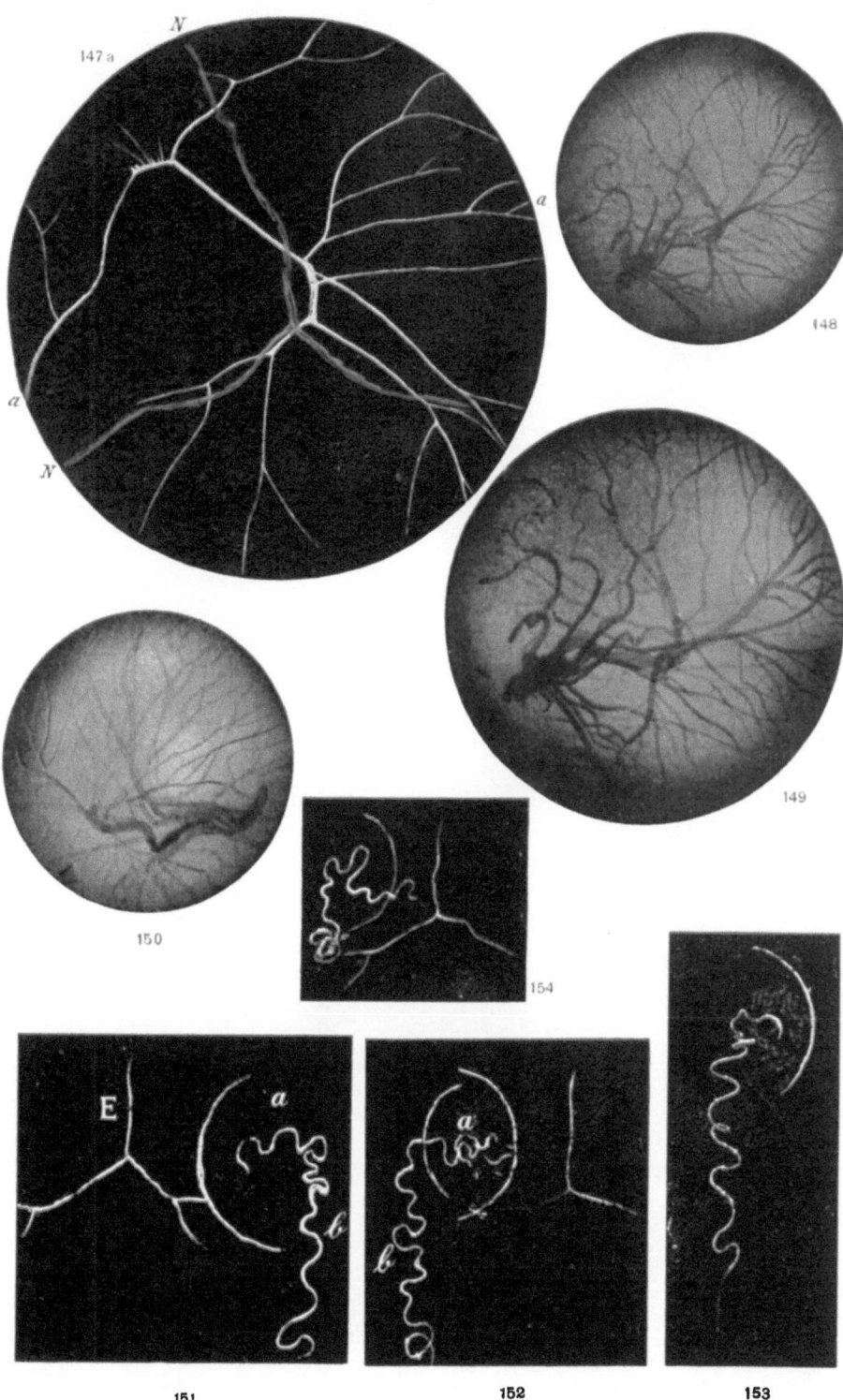

Where is the floating hyaloid inserted? If we examine the area of attachment just described and look in a nasalward direction (Figs. 151 and 152), we see a vessel leaving the skein-like mass in an upward direction and curving convexly. Also note Figs. 148 to 150 wherein we see a similar curved vessel joining the vertical floating section. We rarely find a vessel leaving the skein in a downward direction. It may show an abnormality in the respect that it will run $\frac{1}{2}$ to 1 mm in any one direction along the capsular surface and then enter the vitreous to become a floating remnant. The width of these vessel remnants is very uniform, about 0,05 mm. They often seem flattened. The colour is gray to white, and the floating remnant is more conspicuous than the area of attachment, because of the fact that the vitreous reflects less light, than the posterior capsular surface. One may occasionally note white spots on the vessels, probably histological changes. More rarely they may be deposits.

The slight elongation of the spiral or cork-screw (Fig. 153) shaped free end on motion is due to the fact that the vessel remnant has a free ending in the vitreous. On comparatively rare occasions I have found these vessel remnants to be *fixed* in the vitreous. They are then less curved, but may show a zig-zag shape. In one case the fixed vessel was directed upward. Figs. 150 and 160 show what forms *extensive fixed remnants* may take. The floating remnant may be several millimeters in length, and it may be impossible to distinctly see its end in the deeper vitreous. At one time I observed a large floating remnant of this kind with the ophthalmoscope, it ended in a ragged flake lying on the supporting structure of the vitreous.

While the *area of attachment* is easily discernible in almost all individuals, the floating remnant is not as frequently seen. However its presence can be proved in many healthy eyes. In differential diagnosis, we may at times see similar suspended fibre structures in the anterior vitreous. They are dichotomously branched fibres of vessels, more distinctly seen in age, probably belonging to the early embryonic period (3rd to 4th month), at which time the vitreous shows a rich arterial vascularization. (See Fig. 332a and b.) (Peripheral vascular area of the vitreous — *Vasa hyaloidea propria* Arnold[56]), Kessler[65]), H. Virchow[66]), Köllicker[67]), Schultze[68].) They are of very small calibre, irregular in contour, and the fibres may cross one another. (Fig. 164 shows fibres of this kind in a man aged 60. Fig. 165 in a girl age 17.) As a rule the fibres illustrated in Fig. 164—165 show irregular white deposits, which makes them appear thickened. I n other cases they at times split or show branching. At times they swing freely, again they seem stretched. They may resemble fibre remnants of the pupillary membrane. They are more numerous and better visible in the periphery than in the axial area. I hardly ever fail to find these fibres in eyes with clear media. Often they are separated from the lens or the retro-lental space by a uniform layer of the supporting structure of the vitreous.

These fibre formations cannot be confounded with the arteria hyaloidea, on account of the typic and constantly uniform location of the latter, as well as on account of their special and peculiar form. We do not advise the use of a magnification higher than 24—37 times (Oc. 2, Obj. a2 or a3), as the spontaneous motion of the eyeball would interfere with correct localization. The pupil of the patient must be fully dilated. The narrow area of the bundle of light must fix the posterior capsule, in order that we be able to observe the area of attachment of the hyaloid. Floating remnants may be observed through a narrow pupil.

Figs. 166 to 178. The white curved line (normal) surrounding the postembryonic hyaloid attachment.

This white curved line, which we have recently described for the first time[69]) can be found in most normal eyes. It is, as a rule, more distinctly seen temporal-ward of the hyaloid attachment. Quite rarely it may completely surround the whole attachment (Fig. 171).

The most important forms of this curved line are shown in Figs. 166 to 178. For the purpose of localization, the embryonic suture is also shown. The central axis of the latter is approximately in the antero-posterior optical axis of the lens.

To observe the curved line a magnification of 24 times is sufficient. (Oc. 2, Obj. a2.) The light is from the temporal side, the patient looks slightly nasalward. If the curved line is on the reflecting zone, the patient must be directed to look farther nasalward, or nasal and up or downward, which displaces the reflecting zone in a corresponding direction.

The line curved in a regular manner is 0,02 to 0,06 mm wide. It usually passes midway between the hyaloid attachment and the posterior pole. It is most uniform and regular at this place. The radius of curvature of the line at this point is about 1 mm. If the reflecting zone covers this curved line, the latter may be seen as a dark fibrous stripe in luminous surroundings. As the reflecting zone is brought about by a reflection of the posterior capsular surface, it proves that the curved line is probably *on the capsular surface*. It may show a decided asbestos-like radiance. We determine the approximate distance of this line from the hyaloid attachment in a similar manner as we have determined the distance of the latter from the posterior pole. Its greatest convexity is about 0,75 to 1 mm from the posterior pole.

At times it may be nearer to it or farther away, or the hyaloid attachment may connect with the curved line as in Fig. 166. Often it surrounds the attachment of the hyaloid (see chapter "Arteria hyaloidea" and Figs. 166, 169, 170), although, as a rule, the upper and lower ends end as in Fig. 167. The vertical distance between these two ends is $1\frac{1}{2}$ to 2 mm.

At times the ends continue in a straight line (in a parabolic manner), or they may curve spirally inward (Fig. 168), or downward (Fig. 169). The curved line often shows thin irregular fibre-like stripes extending toward the hyaloid attach-ment, which in all probability present vessel remnants (Fig. 166).

The temporal zone is often surrounded by similar concentric parallel lines. These thin luminous fibre-like stripes are shown in Figs. 166 and 171. They may extend to the posterior polar zone and even beyond. They are only visible in in-tense illumination, and *they contribute to the greater luminosity of the surrounding area.* (I found this condition in a similar manner in full grown rabbits and dogs.)

This increased luminosity creates the impression that the curved line presents a distinct thickened border of a thin grayish membrane deposited over the posterior surface of the lens.

Could the line represent the everted capsular wall of the hyaloideal canal?

If this conception be true, the hyaloid artery would be inserted in a non-capsu-lated area on the posterior capsule. It should be mentioned that *Seefelder*, based on his investigations, denies the existence of a hyaloid canal in the fetus.

In addition to the curved line just described there may be two other structures of similar appearance situated at two different places. In the first place there may

Fig. 155—166.   Tafel 18.

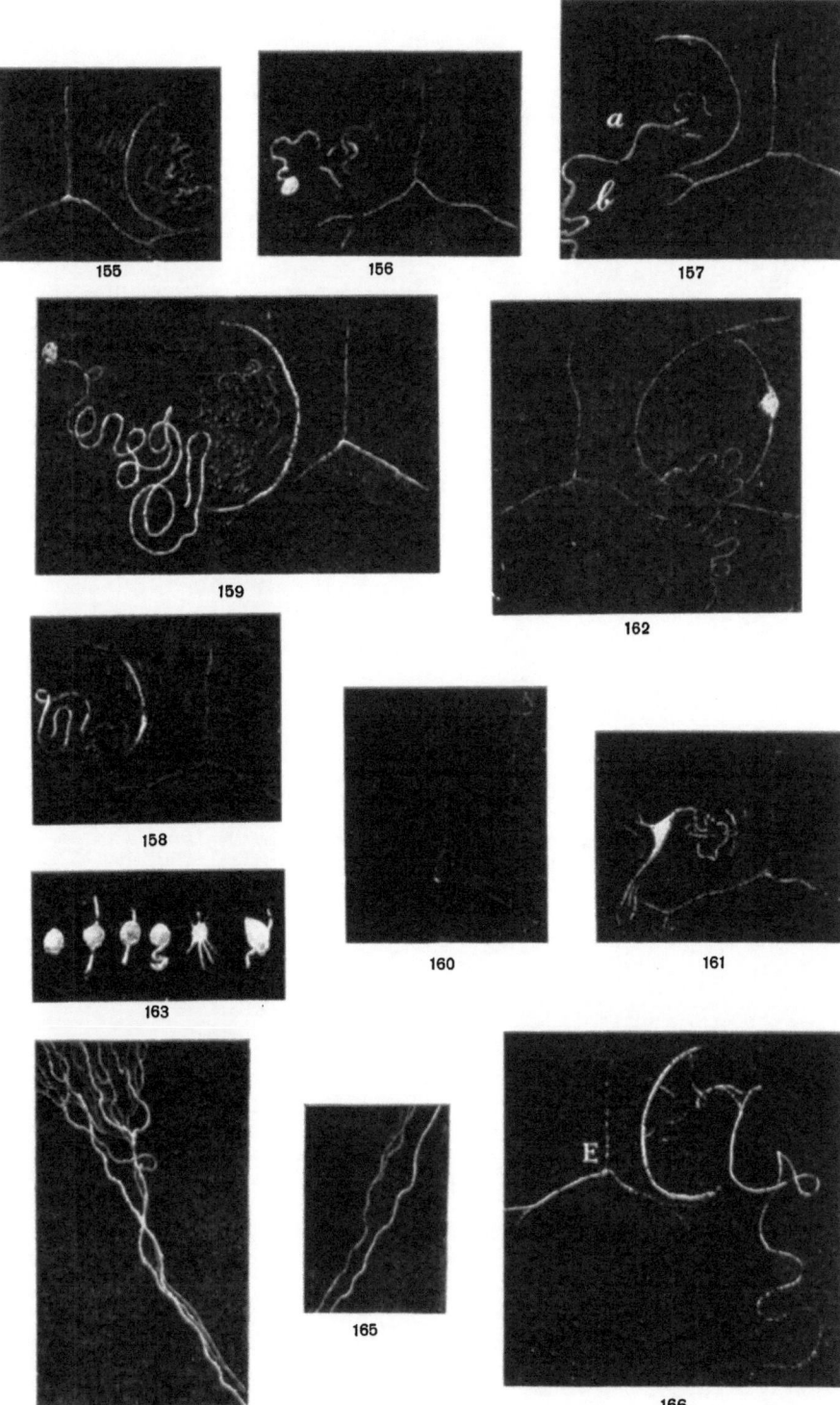

be nasalward of the curved line, a similar second line curved in the opposite direction (Fig. 170).

This and the original curved line may unite above and below, forming at each junction an acute angle. This second line is usually shorter and less luminous than the first line and surrounds the hyaloid attachment (Figs. 170, 171). Very rarely it joins the first line in such a manner that the two form a circle (Fig. 171).

In examining we must again repeat, not to cover this line by the reflecting zone. Let the patient as has been described look in a more nasalward direction.

In the second place there are cases in which there is no curved line visible at all, but one may note the concentric fine fibre striping which has been described (Fig. 172).

The hyaloid remnants as well as these curved lines, owing to the greatly increased internal lens reflection, are more difficult of observation in older individuals. Therefore it is advisable to make observations of these structures on youthful individuals, and not neglect to bring about a fair dark adaptation in the observer's eyes. In 200 cases examined, I found only two cases showing an odd relation of these curved lines. The appearance of these lines in the four individual eyes is illustrated in Figs. 174 to 177. The lines show the typic form and location, but are *divided into segments*, which latter show a *radial arrangement*. The radial segments are especially distinct if we reduce the angle between the line of observation and the axis of illumination. In one of the cases (Fig. 176) the second curved line joins the first, in a manner so that they together form a nearly complete circle.

Just as in humans, I have found the curved line similar in form and extent in full grown rabbits (Fig. 178), also in dogs. On account of the increased antero-posterior lens diameter in these animals, it was necessary to iridectomize or to cut the cornea and iris off of the eyes of dead animals, in order to make these observations. In rabbits, the curved line is *not nasalward* from the posterior pole, but to the *temporal side*, and its concavity is directed to the *temporal side*. Here we also find in the centre of the curved line the relatively dark area of skein-like hyaloid remnants. Especially often do I find the inward eversion of the lower edge in rabbits as shown in Fig. 168. Also on the side toward the suture I find the concentric fibre formation (Fig. 178).

In dogs I also found the curved line to concave temporalward. If we realize the low state of development which the lenses of dogs and especially of rabbits occupy among mammalia, one must come to the conclusion that the observations just recorded have a distinct value in the study of genesis.

In the rabbit, as I have determined by examining 30 living and dead animals (in the latter by removing cornea and iris—leaving the bulb in situ), the posterior suture is directed *diagonally horizontal from the anterior upper nasal side to the posterior lower temporal side.* (In the literature one usually finds an opposite statement.)

Usually the angle is one of 20° to 30° off of the horizontal. Rarely the suture is horizontal, at times it is "S" shaped or curved.

The *anterior suture* does not show the characteristic kinking as Fig. 178 shows, of the *posterior suture*. It is directed at right angles to the latter. Albinos are not as useful for these observations as are normal animals. The distinct outlines of the lens picture suffers on account of the diffuse reflection from the fundus. The fact that the curved line surrounds a seemingly darker zone (outside of which there may be concentric striping), makes me think that the curved line represents an *eversion*, and it is in all probability the distal end of the (embryonic) *hyaloid canal*. Especially convincing are the studies of similar lines in the eyes of rabbits, dogs, and other mammals, where the line is often more distinctly seen, and its similarity to an everted thin membrane is quite evident. This latter covers the posterior lens surface and is bent dorsalward in the area of the hyaloid artery, whence it would form the wall of (capsule) the vitreous canal of the hyaloid artery. In which manner the rare occurence of segmentation of this line can be co-related to this assumption must be learned from further study.

2. THE PATHOLOGIC LENS

Certain changes I have noted occupy the borderline between normal and pathologic lenses. If they are present only to a minor degree, we must regard them as being normal, because they are manifested in a great number of individuals.

This class comprises:

1. Individual or a limited number of punctate or lineal opacities.

2. Iridescence of slight degree in the anterior and posterior reflecting zone (posterior pole).

3. Individual spherical bodies of the anterior graining (shagreen).

Fig. 179. Individual circumscribed punctate and lineal opacities (microscopic punctate and fibrous clouding of the cortex).

T. W., age 11. A clouding 0,08 mm. in size within the posterior lower layer of separation, further upward a small dot. (Narrow slit.) These changes are found in all normal and pathologic lenses. The small dots are mostly found in the periphery of the lens, in the deep cortex or outer nuclear areas. They are from 20 to 100 microns in size, quite dense (white), sometimes too minute and dust-like to be measured microscopically under high magnification. As these punctate cloudings are present, one or a few, in all lenses, they must be considered normal (Fig. 147). However it is to be noted that they are found increased in numbers and size in age.

A limited number of lineal opacities are commonly found. They are arranged parallel to the fibres of the respective layers and usually in the deep peripheral cortex. Therefore we often find them hook-formed near the equator of the nucleus (Fig. 182). They may also be composed of lines formed by individual dots.

Fig. 180a and b. Faint iridescence of the anterior reflecting zone.

By focussing the anterior graining of the lens in middle aged and older individuals [24] [26]), one can note reddish, often faintly green tints in the luminous reflecting zones (Fig. 180a). That we are not dealing with chromatic aberration or diffraction produced by the illuminating apparatus, can be proved by substituting a white porcelain surface for the eye under observation. This plate will now show, when using the Nernst-fibre, that the colours produced by aberrant rays in the apparatus are located at a different place, hence are not identical with those seen in the shagreen or lens surface.

More pronounced iridescent colours may be seen in persons of these ages on the surface of the posterior lens graining (Fig. 180b). They are usually situated only in the posterior polar area, and are smaller, owing to the reduced curvature of the posterior lens surface. It seems to me that this physiologic iridescence may be due to delicate remnants of the embryonal capsular membrane.

Fig. 167—178.

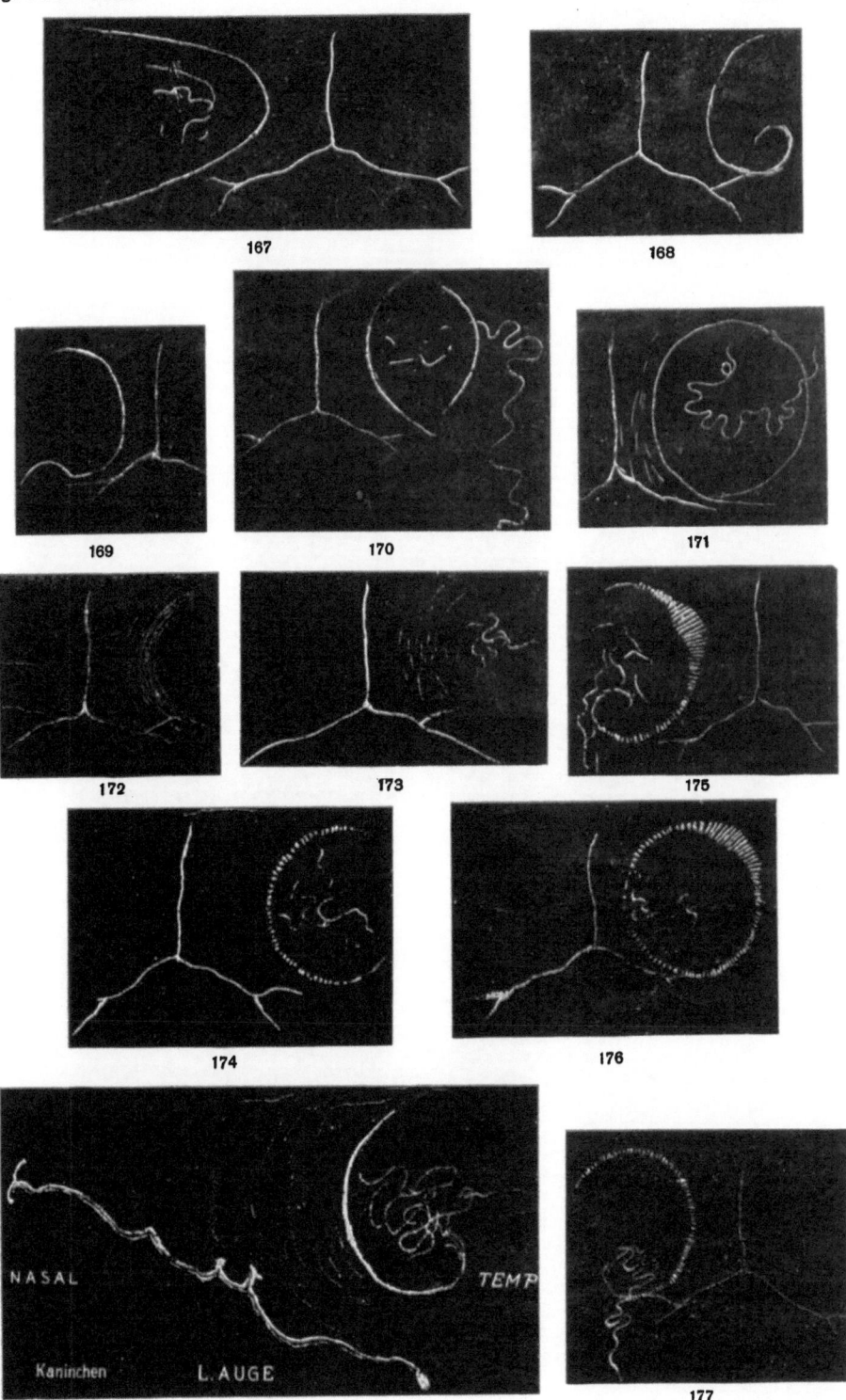

167

168

169

170

171

172

173

175

174

176

178

177

Verlag von Julius Springer, Berlin.

Fig. 92 b (to left and below). Individual shagreen spheres in the area of the anterior reflecting zone.

We describe as *shagreen spheres* structures of a round or nearly round form[22]), about 20 to 60 microns in diameter, situated in the anterior lens graining (shagreen) and only visible by focussing the latter. We have called them *shagreen spheres* which term however does not define their nature. They are usually found in a peripheral zone, corresponding to the junction of the middle and outer third of the radius of the lens, visible in a pupil dilated to the maximum (Fig. 92b). They may be distributed at times in the axial lens area.

Fig. 92 shows individual spheres of this character in a patient 25 years old. They are rarely seen in children, more often in age and especially in cataract (compare Schürmann[23]).

By focussing the graining, these spheres appear as black holes punched out of the luminous field. By indirect lateral illumination, these apparent holes show a distinct mirror-like reflection, which latter reveals their spheroidal shape. These structures cannot be seen by focal illumination with the ophthalmoscope. To observe their borders most distinctly one must use a wide luminous bundle, of a somewhat reduced intensity of light. This veils the more minute changes of the graining, because of the reflection of the capsular surface. In contrast to these three described conditions, which may be considered physiological, owing to their constant occurrence, the following changes must be accepted as pathologic.

Fig. 181. Numerous senile punctate, dust and snow-flake-like opacities of the lens, especially of the deep peripheral cortex. Mrs. J. age 72.

Oc. 2, Obj. a2. Opacities of this nature are found in *most senile, but rarely in lenses of youthful individuals.* They are a symptom of senile or pre-senile lens changes[5]), and represent a transitional stage, leading to a reduced translucency of a sufficient degree to lessen visual acuity. They are, as a rule, situated in the deep cortical periphery, near the equator of the nucleus. Later on they appear increased in number in the axial zone and are more superficially situated, this also tending to reduce the visual acuity. However these cloudings must be quite dense and numerous to materially interfere with central vision.

The punctate and dustlike opacities are found in combination with the various types of senile cataract. In advanced cataract they are practically always present, and often help to cause the peripheral concentric lamellar clouding (see these). Early changes of the punctate type, if not too fine (dustlike), are characterized by an *especial white colouration.* They are dot to fibre-like in form, in the latter case parallel to the lens fibre (see Fig. 182), and are a part of the nuclear zone.

Before the introduction of the slitlamp, this type of opacification could only be observed with great difficulty, often not at all.

Fig. 182. Numerous (juvenile) punctate and hookshaped cloudings.

They are of an intensive white colour, especially in the equatorial area of the nucleus. They are stationary and visible throughout life, in contrast to the opacities shown in Fig. 181.

Figs. 183 and 184. Marked iridescence in the anterior lens reflecting zone (Vogt[24]).

This is found in traumatic cataract, especially following anterior concussion, without perforation of the capsule, or where the latter may have been opened, and has again closed.

This iridescence may also be associated with certain types of senile cataract, especially in the advanced stages.

In my first publication describing iridescence of the anterior lens graining, I reported only red and green colours as being clearly visible. Further observations[26]) have shown other variations, such as yellowish green, bluish-green and yellow, probably dispersion (interference) colours, like those seen in thin flakes. Iridescence of the anterior lens is extremely common. Copper-red and green to bluish-green shades predominate. This phenomenon is seen only in the reflecting zone of the anterior area of graining (shagreen). *Purtscher*[25]) and the author have drawn attention to the fact that this iridescence is especially vivid when foreign bodies of copper are in the eye.

Figs. 183 and 184 show this phenomenon as a symptom of contusion. K. W. age 22. Contusion cataract. Two years ago loss of the right eyeball by explosion of a shell. Left traumatic cataract. Vis = $^6/_9$.

Anterior and posterior cortex, axially and peripherally numerous punctate and dustlike opacities. Temporal and up to the present, coloboma of iris (Fig. 183).

Fundus: Temporal and upward, veiled by clouded vitreous, apparently an area of chorioretinitis. No evidences of perforation.

Fig. 183 shows the cataract magnified 10 times. In the centre, clouding of the capsule, surrounding this, striped and netform opacities of the superficial cortex visible by transillumination.

Fig. 184, same case, increased magnification. Oc. 2, Obj. a2. Marked iridescence, visible by focussing the anterior reflecting zone. Anterior to the *dense white capsular opacity*, which latter shows characteristic curved depressions, there is no graining or iridescence. A dark halo surrounds the opacity, also showing an absence of graining. This halo, showing fine folds at the tips of the cataractous areas, is always present in opacification of the capsule. The predominating colours are red and green. To the left the colours appear uniform and as if on a flat surface, to the right they show a marble-like arrangement.

Fig. 185. Iridescense of the anterior lens reflecting zone in advanced senile cataract[24]). *E. B. age 73. Left intumescence.*

The picture shows broad fissures (spokes) containing fluid, at the sutures. Their ends join axially. Thin white lines pass through these spokes connecting the sutures. (See the picture of lamellar separation Fig. 207.)

The iridescence of an area over one of the spokes, which latter extends to the cortical surface, is especially distinct. Oc. 2, Obj. a3. In addition there is nuclear sclerosis. (The right eye was operated on for the removal of cataract two years ago.)

Fig. 186. Geographical, maplike, interruptions of the anterior graining by delicate zones of exudative deposits on the anterior capsule.

I have observed changes of this kind after trauma[11]), with hyphema, also after various types of inflammation. With suitable direction of light, the graining in the exudative zones may appear less distinct or absent.

These deposits are often so delicate in nature that they can only be observed by exactly focussing the graining (shagreen). In this (Fig. 186 the left eye of the seven year old boy R. S. with kerato-globus and quiescent parenchymatous keratitis), and in other cases I have noted iridescence of the thin exudative membrane. Oc. 2, Obj. a2.

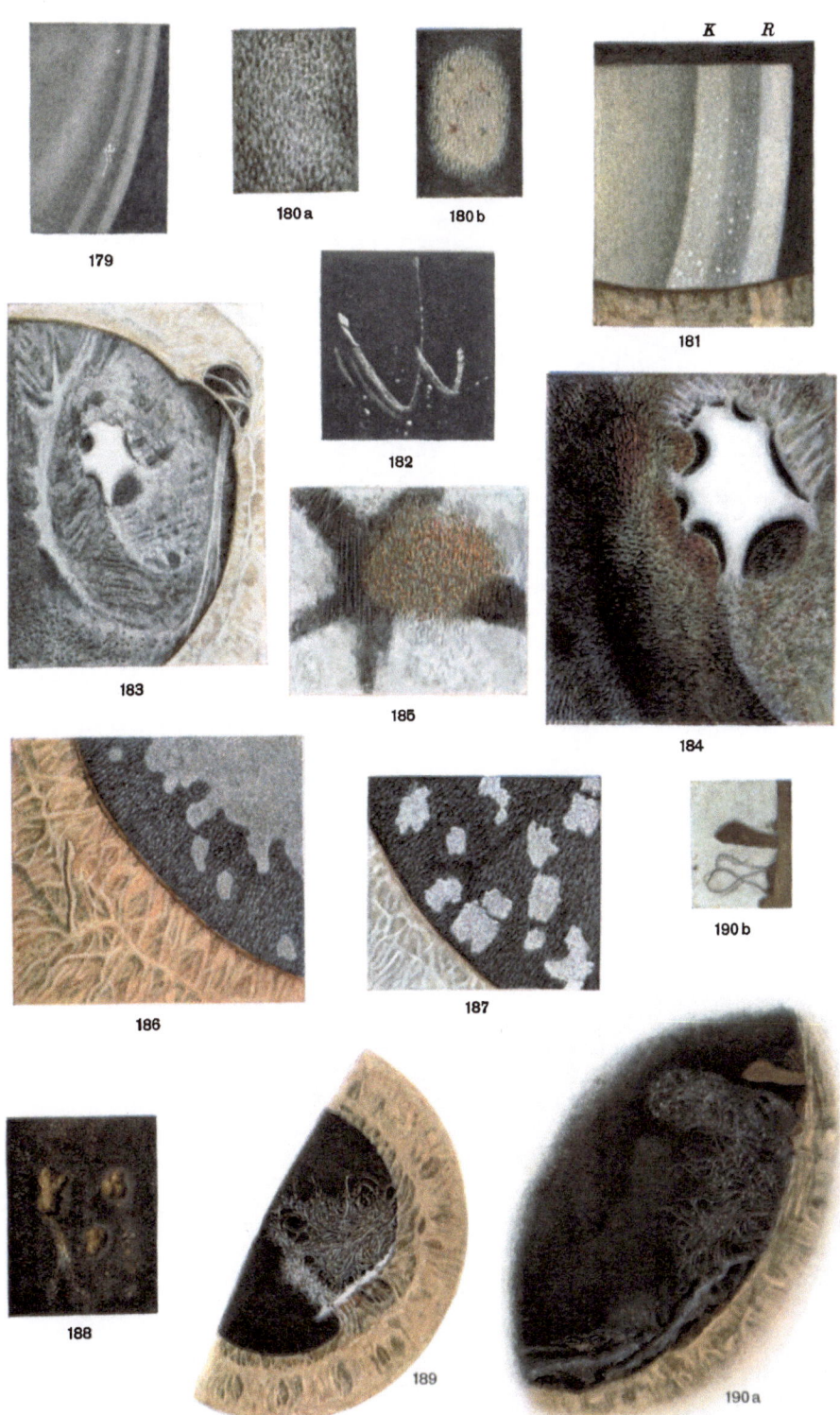

Fig. **179—190**.

Tafel **20**.

K   *R*

179

180a

180b

181

183

182

185

184

186

187

190b

188

189

190a

Fig. 187. Iridescence of exudative remnants on the anterior capsule. (See text to Fig. 186).

Exudative deposits of this nature often develop during the course of an iridocyclitis, on the anterior lens capsule under the iris, and can only be seen when the pupil is dilated to its maximum. This is so in Fig. 187, which shows the anterior lens surface of Miss S., the case of whom is also illustrated in Fig. 316 (iris tubercles).

In areas surrounding the exudates the graining presents a map-like character. Oc. 2, Obj. a3. They often cover quite an extensive area and may be visible for many years after the condition has become quiescent.

Fig. 188. Fresh flake-like deposits of pigment on the anterior lens capsule, following iritis. Mr. E. age 43.

Oc. 2, Obj. a2. The pigment clumps, under high magnification, show a delicate granulation of their surface. A grayish exudate binds them to the lens capsule.

Fig. 189. Exudate and pigment showing delicate fibre-like and star-shaped changes.

H. S. age 42. 12 years after double perforation by a spicule of iron. Low magnification.

Broad posterior synechia, at the site of perforation a diagonally lineal, densely white scar. The capsular exudate has assumed a delicate net-like design, in which starshaped pigment masses are situated. (In addition there is posterior cortical cataract of the cataracta complicata type. The path taken by the spicule of iron through the lens shows as a lineal opacity).

Fig. 190. Exudate and pigment, showing fibre and star-shaped patterns. Mrs. G. age 29. Chronic iridocyclitis.

Oc. 2, Obj. a2. Presenting at "a" to the right and above a pigment strip and below this a more recent, as yet, little changed exudation. By *transillumination* at "b", one can observe two new vessels extending from the iris. Below these the typic, in part, whorl-like star-shaped figures and fibres.

Fig. 191. Numerous "Shagreen spheres" in the anterior reflecting zone[22])[23]). (Compare text page 75.)

They are situated in an intermediate zone, near the junction of the middle and outer third of the lens radius. Certain spheres show a lateral illumination, others appear as dark holes in a luminous area.

Figs. 192 and 193. "Shagreen spheres" under higher magnification.

Oc. 4, Obj. a3. In Fig. 192 the spheres are relatively small and varied in shape. It Fig. 193 they are larger and round.

Fig. 192 shows the shagreen spheres in a case of coronar cataract. Miss C. H. age 65, also iridescence of the anterior graining and incipient spokes. Vis = $^5/_6$: Fig. 193 M. S. age 60 coronar cataract, cystic spaces and spokes. Vis = $^1/_2$.

Figs. 194 to 196. Vivid iridescence at the posterior reflecting zone.

This symptom visible only in the reflecting zone is characteristic of incipient cataracta complicata, however it may at times be observed in other types of cataract. (Compare Vogt[26])[72]). It is found almost without exception in the posterior polar area. The fields of varied colours (yellow, green, blue, red) are unequal in size and irregular in shape, arranged in a maplike manner. In ordinary focal (diffuse) light, one sees instead of the iridescence only a thin cloudlike film (Fig. 196) in the subcapsular cortex. The colour zone has often an irregular form and may show radiations. In advanced cases of cataracta complicata, I have found this iridescence not alone in the axial area, but also in the periphery of the lens[72].

Fig. 194. Iridescence of the axial posterior reflecting zone in a case of cataracta complicata incipiens. M. A. age 42. Emmetrope.

Oc. 2, Obj. a3. The vision of his left eye was lost on account of post-traumatic amotio retinae. Anterior and posterior reflecting zones show iridescence, *the posterior only in the axial area* (Fig. 194).

Anterior and posterior cortex in the superficial axial, in part in the deeper zones, show a faint clouding. In the posterior cortex a large layer of subcapsular vacuoles. (Shown in the lower part of the figure.)

The opacities are not visible by focal illumination with the ophthalmoscope.

Otherwise the lens shows no differences from the one of the (normal) right eye, the plastic image of the nucleus and the embryonic sutures are somewhat more pronounced.

Figs. 195 and 196. Maplike iridescent area in the axial zone in cataracta complicata.

Mrs. M. B. arterio-sclerosis, has been under observation for six months on account of retinal hemorrhages. (R Vis = $^1/_3$. Three months after the pictures were taken she suffered an apoplectic stroke.)

The right posterior lens pole, under a 24 times magnification, shows the subcapsular maplike nebular clouding, as seen in Fig. 196. By focussing the reflecting zone a vivid iridescence shows in this area (Fig. 195). Outside of this area there are slight nebulae in the posterior cortex, seen with some difficulty.

Figs. 197 to 205. Cataracta traumatica.

(In reference to iridescence in traumatic cataract see Figs. 183 and 184.)

Figs. 197a and b. Cataracta traumatica intumescens.

Railroad employee E. G. age 40. Perforation by iron splinter four weeks ago. Oc. 2, Obj. a2.

The large flat splinter is imbedded in the anterior capsule and cortex (Fig. "b"), opposite the corneal scar (wound of entry). It projects slightly into the aqueous. The lens shows a bluish white clouding. The anterior chamber is slightly shallow. The superficial fibre bundles of the lens show an asbestos-like gloss. The cystoid spaces (fissures) between the fibres and in the areas of the sutures are quite broad. *The bundles of lens fibres* are separated by these *cystoid spaces*, so that individual fibre bundles pass diagonally through them, these latter show in a manner similar to wood splinters connecting two halves of a split board. From this one must infer that the fluid when being absorbed by the lens must exert a certain amount of pressure. (I have noted similar appearances in senile intumescent cataract.)

Fig. 198. Traumatic cataract by transillumination.

Z. T. age 31. (Spicule of iron perforating 14 days ago, discission 5 days ago.) The illustration shows the upper outer cortical zone by transillumination. Note the tubular vacuoles between and along the fibres in the upper part of the illustration, showing a featherlike configuration, leading toward a suture. In the lower part of the illustration, there are irregular subcapsular *amorphous white areas*, among them perfectly round vacuoles.

The amorphous areas are probably the results of a junction of vacuoles, as they seem to be composed of clear fluid. They are therefore luminous by transillumination. The round vacuoles remind one of the "myelin droplets" as seen in senile cataract. (Morgagnian globules.)

Fig. 199 a and b. Traumatic Cataract in focal light.

Fig. 199a shows the case of **Fig. 198** *in focal illumination.*

The character of the luminosity is now just the opposite of that in Fig. 198.

The amorphous areas, as well as the vacuoles in the lower part of the illustration now appear dark with lighter borders, just the reverse of the manner as shown in Fig. 198.

Fig. 199b shows the anterior lens surface in low magnification (8 times), 9 days after the discission and just preceding the lineal extraction. Focal illumination. The white triangular indistinct area is the site of perforation by the spicule of iron. The discission wounds of the capsule are not very distinct, the whole lens substance has become slightly opaque and the sutures and fibre design are quite evident.

Fig. 200. Old stationary circumscribed traumatic cataract.

K. age 29, circumscribed opacity in temporal cortex. Eleven years ago he sustained a double perforation of the lens by a spicule of iron. A corneal, iris, anterior capsular and double lineal lens scar (see Fig. 200), the latter diagonally central, causing a round dense opacity. Oc. 2, Obj. a2.

The opacity of the posterior cortex extends to the subcapsular area, is densely white, porous and suggests a complicated cataract. It involves the whole cortex and extends into the nucleus, its borders contrast with the clear lens zones by the interposition of a slight nebular clouding. The double lines to the left show the direction taken by the foreign body on its way through the lens. They show brownish red pigment debris, which may be derived from blood, iris pigment, or from the foreign body. Some of the small dots are luminous. No iridescence, no siderosis, no posterior bulbar wound of exit visible. Vis $= \frac{1}{3}$.

Fig. 201. "Vossius' Ringtrübung".

(*Vossius'* post-traumatic annular opacity, more correctly ring deposit on the anterior capsule.)

Hesse[73]), and soon thereafter the author[11]), have shown that this socalled circular clouding is not a clouding of the lens substance or the anterior capsule, but is a thin deposit on the external surface of the latter. Whether this deposit is composed only of blood as assumed by *Hesse*, or whether it is composed mainly of fuscin-granules and pigment of the iris border, as presumed by the author, can only be definitely proved by anatomical investigation.

I found the deposit so dense in two cases that it veiled the anterior lens graining by virtue of its dense dark ring. (Similar though maplike irregular interruptions of the graining are caused by delicate exudative membranes on the anterior lens capsule. See Figs. 186 and 187.)

Figs. 202 a and b. Traumatic rosette shaped posterior subcapsular cataract and its resorption (compare Vogt[72]).

On the evening of January 14th, a 16 year old girl perforated her right cornea iris and lens with a needle. The needle entered the cornea 2 mm from the temporal limbus and penetrated into the vitreous. At the first examination 36 hours after the injury, there was a slightly infiltrated corneal wound and a typic posterior rosette shaped cataract. R Vis. = $^6/_{18}$, L Vis. = $^6/_6$.

As can be observed in Fig. 202a (taken January 19th) a horizontal lineal opacity about 1 mm in height, showing a fibre design extends backward to the temporal posterior capsular wound of exit. At the site of the latter, on the left side of the illustration, there is an irregular circumscribed perforation cataract. In this area there is also a flattened vacuole about 1 mm in diameter, visible by transillumination. (This vacuole disappeared after a few days.)

The cataract has a round rosette-shaped border and a vertical diameter of 4,5 mm (January 16th to 20th). The Slitlamp and corneal microscope disclose seams and fibre-like designing in its borders. In the fringed area the opacity is less dense, is composed of minute luminous dots, which seem to surround the lens fibres and certainly *define their outline.* The whole clouded area (as is also seen by transillumination), is uniformly thin. Owing to its superficial luminosity it reminds of a layer of small boric acid crystals.

The "reflecting zone", when brought to the border of the cataract, shows the latter to end abruptly, because it is covered by the capsule. On the opacity it shows as a luminous mat reflex, diffusely disseminated into the surroundings. No iridescense was noted in the opacity, nor at its borders. At the border of the cataract one can distinctly note the separation of the *layer of cataractous droplets* and *the embryonic deposits on the posterior capsule.* The capsule separates these two structures by its clear layer of uniform thickness. Here one can also note the extreme thinness of the layer composing the cataract.

The anterior site of entry into the lens is covered by an iris synechia. The direction of the needle through the lens shows as a thin antero-posterior line of opacity. The anterior fibre and suture configuration is conspicuously increased, compared to that on the left (normal) lens.

No iridescence anteriorly. We have examined this cataract daily from the 16th day of January to the 12th day of February (26 days) with the slitlamp and corneal microscope, as well as by focal illumination with the ophthalmoscope. During this period we have proved that the posterior traumatic rosette shaped cataract is subject to almost complete resorption. (Compare the corresponding findings by E. Fuchs[73]) who in all probability first drew attention to this fact.)

As early as on January 21th we noted a clearing up at the periphery. In the following days one could observe, step by step, the resorption of this rosette-shaped cataract. In this case it has, with the exception of a very small area, disappeared inside of 31 days. (Six weeks later this small change was completely absorbed with the exception of a very few minute dots. Vis. = $^6/_6$.)

Originally it had a diameter of 4,5 mm, on the 10th day one of 2 mm, 10 days later one of 1 mm, and at the expiration of 10 more days was reduced to a diameter of 0,75 mm. In the latter stage the remaining opacity was composed of a few fine dots, barely visible by focal illumination with the ophthalmoscope. The visual acuity increased from $^6/_{18}$ to $^6/_8$. The posterior capsular scar and the small zone of opacity

Fig. 191—200.   Tafel 21.

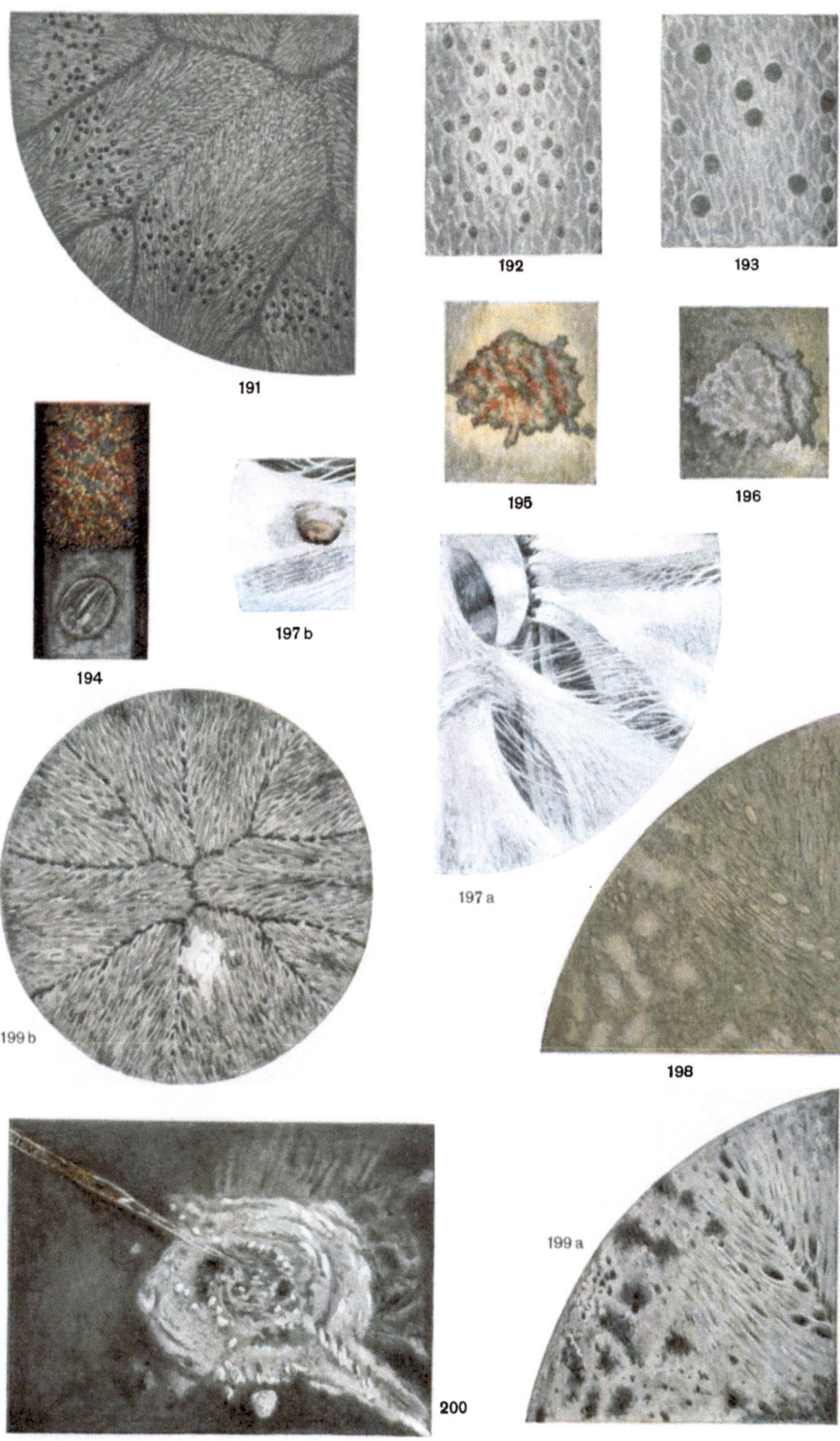

191

192

193

194

195

196

197 b

197 a

199 b

198

199 a

200

Verlag von Julius Springer, Berlin.

Fig. **201—205.** Tafel **22.**

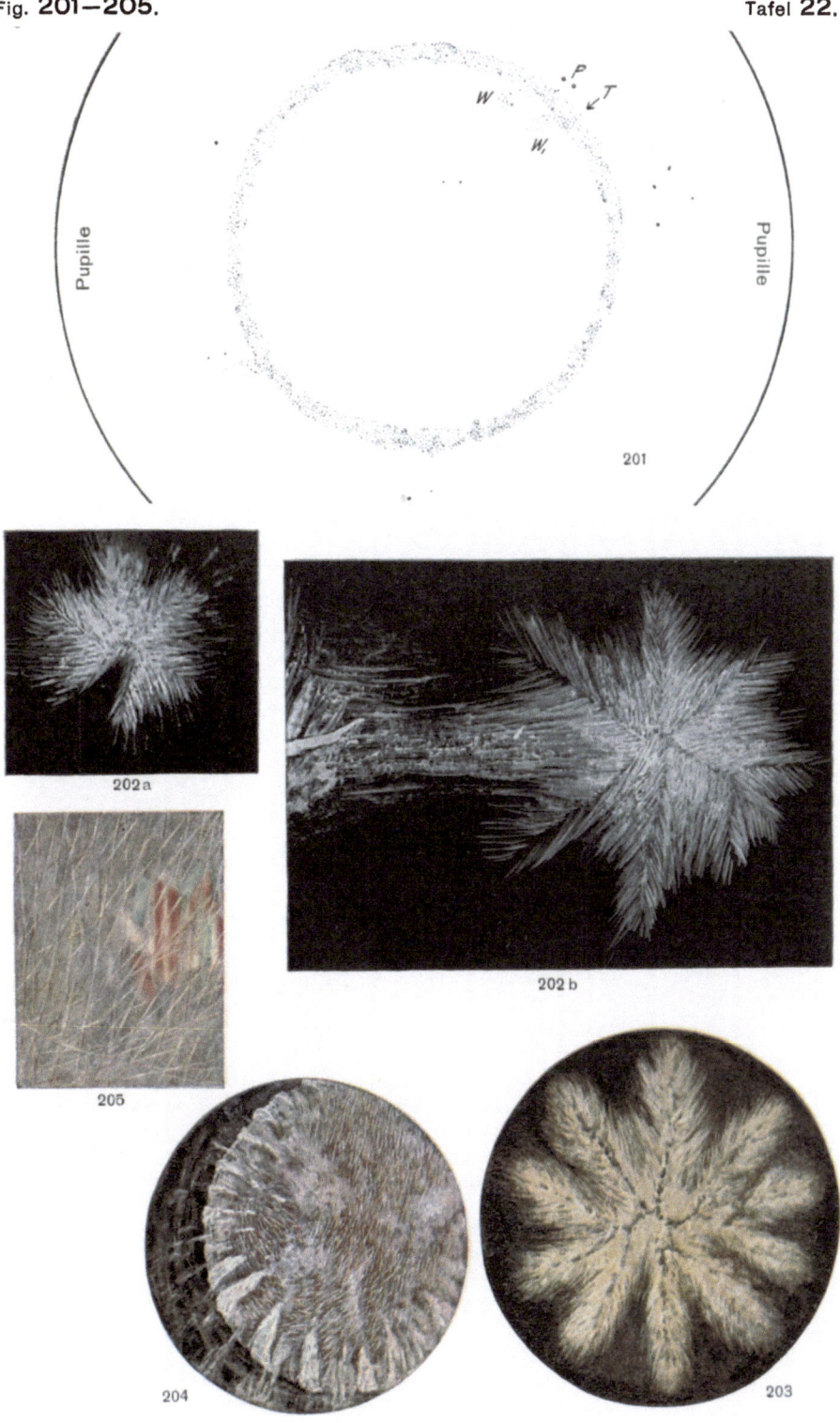

surrounding it will in all probability remain as the only lens changes of a permanent nature. The absorption of the cataract commenced with the peripheral zones. Fig. 202 b shows the opacity ten days after the first observation.

We observed that the clouding was composed of *minute droplets*, which were situated under the capsule and in all probability also between the fibres. The latter is quite likely the case, as the droplets accentuated the picture of the fibre and seam (suture) designing of the lens, in a manner similar to the action of nitrate of silver on freshly enucleated lenses. In the latter case the cement substance between the fibres is darkened.

The differential diagnosis between senile posterior cortical cataract, as well as typic cataracta complicata, and this type of rosette-shaped traumatic posterior opacity is manifested by the extremely thin layer of opacity and the delicacy with which fibre and seam designing of the cortical surface is accentuated in the last named form of cataract. Its site is in the area of the posterior pole, even though the injury may have been at the equator or in the anterior part of the lens.

Fig. 203. Traumatic posterior rosette cataract in a case of perforation by an iron spicule.

Case of Figs. 198 and 199 observed one day after injury to the anterior cortex. Low magnification. Regarding this type of cataract, compare the text to Fig. 202.

In this case, owing to the severe injury to the lens capsule, a needling followed by extraction was done a few days after the picture was taken.

Fig. 204. Cholesterin flakes and luminous (iridescent) fibres in the nucleus. Right eye of F. S. age 9.

$4^1/_2$ years after shotgun injury. Low magnification. Nasally, a small scar at the limbus. It was stated that there was an iris prolapse at this site, at the time of injury. *Extensive concussion cataract*, the nucleus seems displaced upward, and inward within the lens substance (apparently no subluxation).

The outer and lower cortex shows a slight concentric and radial clouding. The nucleus is surrounded at its equator by white linear opacities arranged in a dorso-ventral direction and meeting at a point axially. In the body of the nucleus there are iridescent fibres, like long thin needles, arranged crosswise, parallel and in a tuftlike manner.

One has the impression that they are modified lens fibres. Five months after taking the picture the latter condition was unchanged, however the superficial nuclear opacification had increased somewhat so that the luminous fibres were hidden in part. We now note two rhomboidal, vividly iridescent cholesterin flakes in the upper part of the nucleus (Fig. 205).

R. V. = Fingers at 1 meter. Tension slightly reduced. Anterior chamber, and fundus as much as is possible to observe, are unchanged. Six months later there was complete lens clouding.

Fig. 205. Cholesterin flakes and iridescent threadlike needles. Case of Fig. 204, in increased magnification.

Oc. 4, Obj. a3. Note the vividly iridescent flakes, and the long coloured luminous needles (probably fibres), at various depths, which are arranged in part parallel and again crossing one another. The cholesterin flakes are slightly veiled by faint opacities.

Figs. 206 to 212.   The formation of radial spokes, having fluid contents, in senile cataract.   (Cataracta senilis intumescens.)

The development of spokes filled with fluid is one of the most common changes in senile cataract. They almost exclusively involve the cortex, often in many layers, most commonly in the zone surrounding the nucleus.   By transillumination they appear as dark radiating spokes in the superficial and deep cortex, and usually in the areas of the lens sutures.   Often one may note a splitting of a seam or suture, as has been done experimentally by *macerating a freshly enucleated lens*[46]) [75]).

Fibre bundles separate by maceration in the same manner as in senile and traumatic cataract, in the latter cases the spaces contain fluid (see Fig. 197).

Again as in maceration, myelin droplets (morgagnian globules) enter these spaces. Fig. 207 shows droplets of this kind in focal light.   Fig. 208 shows them by transillumination, whereby their droplet-like nature is manifested.

The droplets bring about a certain degree of opacification, owing to their changed index of refraction.   They present the *incipience of spoke development*[46])[5]).

The latter change, the most common type of lens opacification, is therefore the result of a radial lens fibre and suture separation by fluid.   In a similar manner we may observe the development of lamellar separation in the cortex.   In sclerotic lenses these changes can therefore not be present.   They may precede intumescence, and can always be demonstrated in the latter change.   The fluid which at first separated the lamellae and sutures finally finds its way to the individual lens fibres, myelin droplets are excreted from these, and thereby the condition of intumescence results.

In the more mature state of cataract formation which follows this condition of fissures filled with fluid, the latter begin to disappear, and the volume of the lens, owing to the loss of fluid is probably reduced.

On account of this and other observations (for instance in cataract following massage compare Fig. 210), I have come to the conclusion that the absorption of fluid by the lens is due to internal causes, that is a change within the lens substance. (After massage there is no injury to the capsule or other parts, just an injury to the *inner* lens substance, the living fibres and epithelium, which leads to an absorption of fluid and intumescence.)

When this latter process is ended, by a breaking down of the dying *fibre substance* which had caused the intumescence, the lens will regain its former volume, and the distended capsule will return to its original state of tension.   The state of maturity has been attained.

In contrast to this senile process compare the clinically so different form of *cataracta complicata* which is due to the effect of *exogenous noxious substances*, Figs. 268 to 280.

The radial formation of spokes and lamellar separation are of similar genesis, and usually both changes are present.

The former separates the *sutures*, at times the fibre lamellae of *Rabl*, which latter is a separation of the layers (leaves) of the concentric lamellation of the lens.   Both involve simultaneously structures of a like maturity.   The causative factor is an absorption of fluid by the lens.

Both changes occur in artificial maceration, so that the latter, *to a certain degree*, may be compared to the process of *cataracta intumescens*.

We receive the impression that during the process of senile cataract formation, *lowered vitality* of the lens finally reaches a stage, in which the lens can be compared to a dead lens, under the influence of the fluids which accompany a mechanical maceration. This process may however take many years, in fact decades. I have observed the gradual formation of fluid interspaces in lenses during many years, under careful control by drawings. In some cases progression was noted in a few months, while in others the status of development had hardly changed in years.

We do not know why in one case the stage of rapid progression may be early, that is the vitality of the lens decreases more rapidly, any more than we know why in certain individuals the senile depigmentation of the pupillary pigment border of the iris, or the development of gerontoxon occurs, why the hair becomes gray early in middle life, or for what reason the thousands of senile changes in individual organs or their parts vary so greatly in their time of appearance.

One thing is certain, we are not dealing with *exogenetic noxious substances* in these senile processes, but they are due to changes of life.

Diseases due to exogenetic noxious substances may be avoided or cured. They must not be confounded with the, in principle, different changes due to senility. Senile changes cannot be avoided by *prophylaxis nor therapy*, and most of us are today convinced that there is no elixir of life, or a fountain of youth.

To attempt to explain the typic changes of senility as due to external noxious substances would necessitate an admission of the *existence* of these methods of avoiding old age.

Just as the time of ageing and natural death are characteristic for all species, and in them is transmitted by inheritance, so is also the period of the development of senility a fixed one for the individual and his various organs.

We can prove that more than 90% of all individuals over 60 years of age show senile lens changes of various degrees, just as they show senile changes of all other organs. The question why all lenses do not become completely opaque in age, must be answered as the analogous question, why all individuals do not have gray hair or why they do not develop a complete arcus senilis. One thing we *do know* about the genesis of all senile changes, is the fact of their inheritance. This influence can be proved in cataract development in a great number of cases (Vogt[5]).

Fig. 206. Presenile development of fissures filled with fluid in the anterior cortex (see text preceding this).

We here present a very characteristic case of the development of fissures filled with fluid, especially in the suture areas. Right eye-focal illumination, low magnification, dilated pupil. (Mrs. H. 42 years. Vis = $^6/_{60}$.)

The suture edges are spread apart in a lancet and bag-like manner especially in their centres. In this area of the sutures there are radial stripes of opacification, mostly vacuoles, showing a beginning of spoke formation. Below there is visible a slitlike space or fissure due to the separation of fibres. Otherwise the direction of the spokes is typic of the arrangement of the sutures.

The examination of the fissures and their borders show them to be situated superficially as well as deeper in the cortex. (Narrow slit of the nitrogen light.) They are especially numerous in the middle and deep layers. Some extend through the whole of the cortex. At the same time there is opacification in the coronar zones, and in the cortex *anterior* to the spaces there is extensive lamellar separation (white parallel lines). The posterior cortex shows similar changes. The anterior chamber is of normal depth. The other eye shows similar, more incipient changes. Vis L = $^6/_{36}$. The patient is well and the urine normal.

Fig. 207 and 208. Senile development of fissures filled with fluid by transillumination and in focal light.

Oc. 2, Obj. a3. Mrs. B. B. age 68 (right eye).

Nuclear cataract, subcapsular a few vacuoles, anterior shagreen shows iridescence, posterior cortex a saucer shaped cataract. Anterior nuclear surface as well as the cortex show beginning lamellar separation. In the cortex there are numerous fissures filled with fluid and vacuolar spheres. (Compare Fig. 207 focal light and Fig. 208 transillumination.) Vision: R = fingers at several feet distance.

Fig. 209. Flat senile fissures filled with fluid, due to separation of the sutures in the anterior cortex.

At some distance from the right anterior capsule there are *cylindrical* rubber-hose-like *diagonally situated vacuoles* visible in *focal light*. Oc. 2, Obj. a2. Mrs. M. B. age 66. (Patient Fig. 287.) R Vis = fingers 1 meter. L Aphakia after extraction. The peculiar shape of the vacuoles is in all probability due to concentric lamellar separation, as the fissures filled with fluid and spokes at times show a diagonal separation. There is nuclear sclerosis and coronar cataract. The latter is visible in the periphery and is brown by transillumination.

Fig. 210. Intumescent senile cataract. Miss K. Sch. age 64.

Six weeks ago there was bilateral incipient nuclear sclerosis and posterior cortical opacification. Soon after preliminary iridectomy of the one eye combined with trituration (massage for the purpose of ripening), intumescent cataract developed. One may observe very large fissures filled with fluid between the sutures and fibres. The fibres are opaque, their bundles show an asbestos-like gloss. The direction of the fibres in relation to the fissures show that the latter are due to fibre bundle separation or distended sutures. Here and there one sees fibre bundles pass diagonally through the fissures, so that the impression is created that the separation is due to forced absorption of fluid. Anterior chamber is somewhat shallow. Evidently the massage presents a severe insult to the lens, especially its fibres, they having lost their resistance on account of age. They were destroyed in the massaged area and subjected to the effects of the absorption of aqueous by the dying zone, just as occurs in maceration experiments.

The whole lens substance was then exposed to the action of large quantities of aqueous.

Fig. 211. - Separation of the fibres of the anterior cortex (atypic development of fissures filled with fluid) advanced cataract. Subcapsular vacuolar layer.

Mrs. Tr. age 69, high myopia (preceding extraction). Descendant of a family with history of myopia and cataract. R. V. = $^1/_{200}$, L. V. = $^5/_{200}$.

Anterior and posterior cortical cataract, the posterior is rosette-shaped. Oc. 2, Obj. a2. R. direct light. L. transillumination (reflected from the deep lens area) in part indirect lateral illumination. On the right side one sees fibre bundles separated in a similar manner as shown in the traumatic cataract, case of Fig. 197. The fibres are very distinct, in part owing to their clouding. Superficially there are many white round spots of 40—80 microns in diameter, near these very dense areas of much smaller spots. These spots are in the immediate vicinity of the epithelium. In direct light one can see that they are "droplets"—a subcapsular vacuolar layer (compare 207 and 208). The peripheral cortex is filled with dot and dust-like opacities.

THE LOCALIZATION OF THE DEPTH OF SPOKES AND FISSURES FILLED WITH FLUID ·IN THE ANTERIOR CORTEX

(Regarding the determination of depth in the lens see the introduction of this chapter, also compare text pertaining to Fig. 277 a and b.)

Fig. 212. E. F. age 63, left eye. Anterior nuclear relief image very distinctly visible, gray spokes and fissures filled with fluid, especially downward. Pigment debris on the anterior (and posterior) capsule following contusio bulbi, 3 weeks ago. (Cataract in the right eye similar to that which is in the left. No connection with the injury.)

By regulating the bundle of light we *accurately focus* the . anterior shagreen stripe "*Ch*" *(a b c d)*. With a 24 times magnification we also see the stripe of the relief image of the nuclear surface "*N*". The illumination is from the temporal side (see arrow). By a slight movement of the bundle of light, a focal area of which is focussed onto the anterior capsule in a horizontal direction, we see in this case the ends and borders of fissures filled with fluid *(W)*, as they appear and disappear in edge *f h.*

We infer therefrom that the fissures are very near the nuclear cortical border *(N)*.

(Su = sutures of the nuclear surface appear dark, K = lens nucleus. Of late we are using a very much narrowed focal bundle of light, about 0.05 mm. in width.) Pigment on the capsule and opacities of the cortex behave differently if they are inside of the bundle of light or to the left of it. In the latter case they are illuminated from *behind*, therefore seen by transillumination. The pigment is ·of a light reddish brown only if seen by direct (focal) illumination *(r)*, that is in the area of the surface of entrance *a b c d.*

To the left of this surface it appears black, as it is illuminated from behind (see Fig.). For a similar reason subcapsular opacities appear *white* as long as they are in the area of *a b c d*, and dark if to the left of the bundle of light, receiving the illumination from the rear. By this means they may also be located. Cortical opacities which are *not* subcapsular are seen dark on a light background outside of the optical section *b d f h* (outside of the bundle of light), providing they are sufficiently illuminated from behind, and white as soon as they enter the bundle of light, which also is of aid in localizing them. By means of the lines "*L*" the area of lamellar separation of the deep cortex is shown.

Figs. 213 to 215. Wreath shaped cataract[77][78][79][80][5] *(coronar cataract formerly in part known as cataracta coerulea or viridis, and in part as cataracta punctata).*

This type of cataract, considered as being quite rarely found by other investigators, is quite frequently seen, according to our experience[80]. They are present in at least 25% of all persons of the ages beyond puberty. Before that age however we rarely found it. (Compare also *Gjessing*[141]). The cataract begins in the periphery in a *wreath-formed-zone*, about at the border of the middle and outer thirds of the lens radius. It is in the deep cortical layer and on the nuclear surface. (In the area of the nuclear equator, somewhat in front of and behind it.)

The opacities are in the form of a very thin layer, concentric with the nuclear surface, in other words the are spread out flatly in the onion-scale like layer. Therefore they are in the concentric lamellar layers, which latter are so plainly seen after maceration of the lens. (These are *not* the radial lamellae described by *Rabl.*)

The fact that the clouded layer is so thin gives origin to the blue to greenish-blue colour. (Compare the colour of an opaque layer as seen by focal light on a dark background, for instance of a blue iris.) The so often distinctly seen greenish tint is due to the normal varnish-like colour ("Lackfarbe") of the lens (compare *Isakowitz*, Z. f. A. **19**. 401, *A. Vogt*[9]). A bluish tint is shown in the opacities of Fig. 214, however they are seen in their complimentary colour (brownish) by transillumination.

If the opacities are dense they will be white in focal light, that is the reflection will be so greatly increased that the domination of the short waves in the reflected light is quite obscured.

The form of coronar cataract is therefore a thin layer-like concentric opacification, and it is easily understood why a clouding of this character, if it is situated in the equator of the nucleus, shows in a *lineal* manner, when viewed from the front. Such concentric lineal opacifications are illustrated in Fig. 213 (above). One can identify the surface or layer form of these lineal opacities, if instead of observing them from the front we examine them from the side. There are however in age real *lineal concentric opacities*, which frequently and in great numbers surround the equator of the nucleus and are in the deep cortex (Fig. 215). Postmortem I saw these fissure-like opacities greatly increase in numbers and extent. Regarding the *form* of the flat opacities of the coronar cataract, in their incipiency we usually find them club-shaped (Fig. 213). Their shape often reminds of certain mushrooms (clavaria). The base of the club is bent or indistinctly lost around the equator, while the round axial end of the club is seen having a distinct border. Often these clubshaped opacities are confluent (Fig. 213). They may be arranged in rows resembling incisor teeth.

In the course of years and decades these coronar cloudings increase quite slowly in numbers, until a more or less complete continuous wreath is formed. Axially other new opacities join the same zone (deep cortex), they are however not club-shaped, but are *round, oblong*, or *ring-shaped* (Fig. 213). These opacities are also, at first, quite thin and translucent, by focal light bluish, later becoming more dense and whiter. When they encroach on the axial zone they may affect visual acuity. However the reduction in visual acuity is not very pronounced, because the coronar opacities are so very thin, and they are therefore invisible or just slightly discernible by transillumination.* For instance the eye represented by Fig. 214 still has a visual acuity of 0,5 in spite of the numerous axial opacities in the anterior as well as in the posterior lens areas. Finally the visual acuity is lessened because of their increase in density and they now are *white* in focal light.

The larger of the round opacities have a surface diameter of $^1/_3$ to $^1/_2$ mm.

Coronar cataracts are often accompanied by other types of cataract. Usually one finds the above mentioned punctate opacifications (Fig. 181), in great numbers and density.

Frequently complicating the coronar cataract in age we find a concentric zonular (layer) clouding (Fig. 215 shows this complication). In advanced cases we regularly find a development of fissures filled with fluid (Fig. 214), which lead to the formation of spokes. The spokes in Fig. 214 are brown because they are seen by transillumination. The statement that the presence of both coronar clouding and fissures or complete cataract is always a coincidence, is not correct according to our experience.

---

* To describe opacities of this nature which owing to their thinness are visible only by focal light, the expression "Scheintrübung" (false or pseudo-clouding) was recently chosen by mistake.

Fig. 206—214.   Tafel 23.

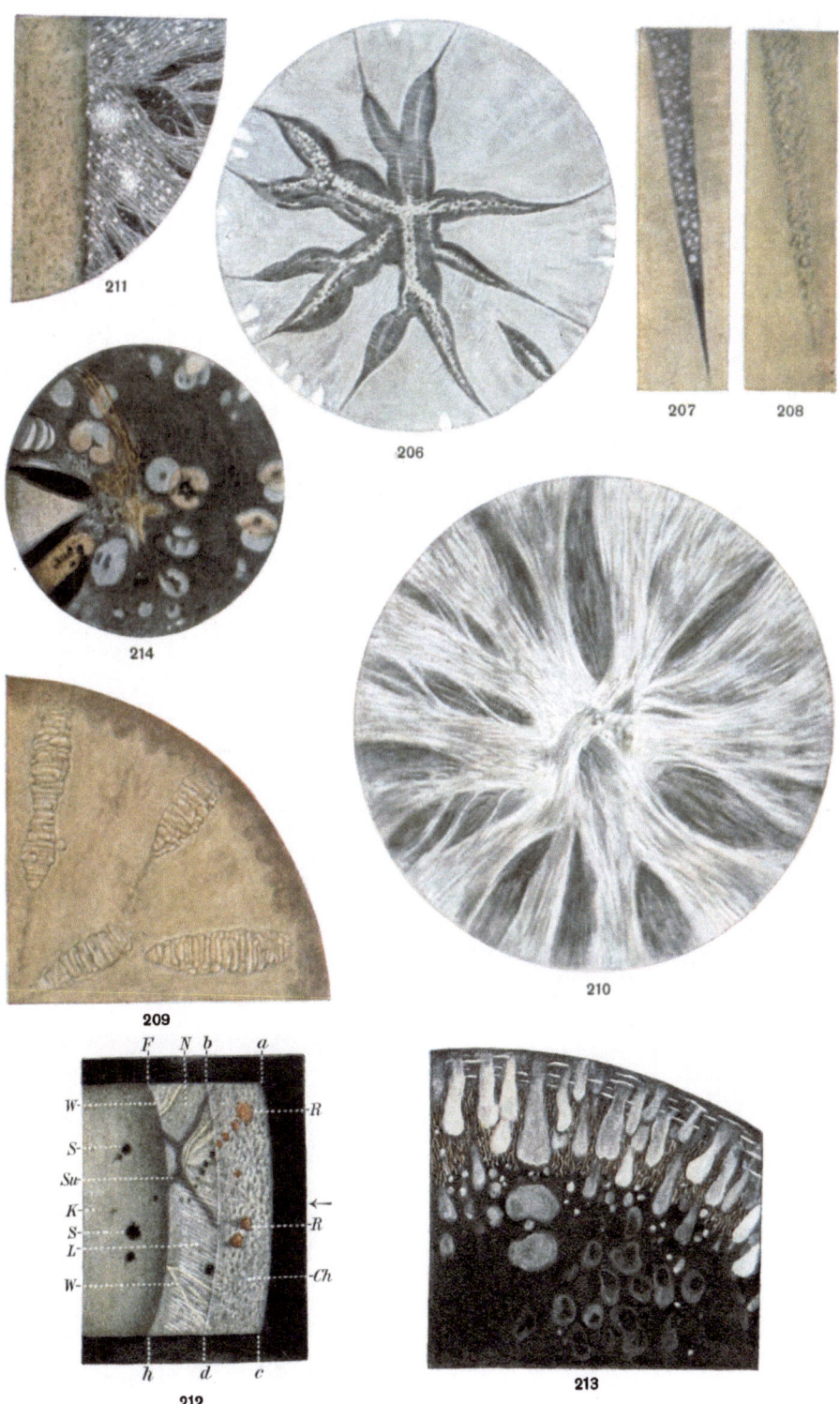

211

206

207   208

214

210

209

212

213

The not very rare combination of coronar clouding and fissures in *young* individuals helps to confirm our assertion. (Compare Fig. 244.) Lamellar separation is also frequently observed.

Coronar cataract as such does not usually lead to complete opacification except in the higher ages. I have found incipient stages of coronar cataract not alone in youthful but in older persons. In a great number of families I was able to prove a marked tendency toward inheritance of coronar cataract [77] [79] [5]). It is rarely found in but one individual of a family group. The presence of low degrees cannot be diagnosed without dilatation of the pupil to its maximum. As this type of cataract is situated in the periphery behind the iris in youth, and does not therefore affect visual acuity we can ascribe to it no value as having an *eliminating* influence biologically. This accounts for its frequency. If the clouding of the lens would contrarily originate axially (central), it would thereby so influence visual acuity early in life that in this manner it would in time tend to an elimination of the strain of individuals afflicted with this highly inheritable lens change, and would not be found with the enormous frequency; vis: 20 % of all individuals.

Fig. 213. Coronar cataract. E. K. age 55.

Oc. 2, Obj. a2. Dilated pupil. In the upper temporal pupillary area there is an anteriorly situated opacity. Note the peripheral flat clubshaped opacities, with their bases bent dorsalward, and the concentric lineal clouding of the deep equatorial cortex covering them.

Among the groups of clubshaped brownish, fibre-like curved opacities, there are axialward round, and in part, ringshaped opaque spots.

Fig. 214. Coronar cataract, progressive axially complicated by fissures filled with fluid and spoke formation. Mrs. E. F. age 54.

Oc. 2, Obj. a2. Right eye. Undilated pupil, about 3 mm. The central blue opacity has a diameter of 0,36 mm. In the deep middle, and superficial cortex, there are fissures filled with fluid, inside of these there are brownish opacities (brownish because seen by transillumination). R. V. = 1/2 with 1,25. Left eye similar.

Fig. 215. Coronar cataract, combined with concentric zonular (layer) opacification. Mrs. G. R. age 61.

Oc. 2, Obj. a2. Right eye. Pupil dilated to maximum. Peripherally the clubshaped and white concentric lines of the coronar cataract. Axially they are joined by the yellowish concentric zonular opacities, which lie in the equatorial nuclear cortical borders. On the anterior capsule one sees fine *radiating* pigment lines, the origin of which cannot be explained. The patient had a mild type of chronic cyclitis a year ago, which may account for their presence. There is however in the area of these pigment lines a slight degree of uveal ectropion (see Fig. 319), which may suggest that there might be a congenital pigment deposit on the anterior capsule. We have found but one case of this kind of pigment deposit in the literature (Brückner [60]). The remnants were similar to those in our case, however there was no ectropion. Brückner correctly assumes them as being derived from the membrana capsulo-pupillaris.

Figs. 216 and 217. Flat wedge-shaped peripheral lens clouding.

I have frequently found this type of opacification [5]) (Figs. 216, 217). It spreads in a similar manner as coronar cataract, that is in concentric flat layers of the deep and middle cortical periphery, which illustrates the onion-like layers of the lens development, just as we see it shown in maceration and in the coronar type. The opaque flat surfaces usually have an abrupt end axially, at times rounded and again as pointed wedges.

Often they extend onward and form spokes. This type of opacification is found in the anterior as well as in the posterior cortex. Anterior and posterior areas may join at the equator. At times however both areas end equatorially in a regular concentric line.

This wedge-shaped peripheral clouding is found as a rule only in old age, just as the coronar cataract is characteristic of youth and middle life. They are thicker and more dense, and by focal light therefore show white to yellowish-white. (The latter especially, when deeply situated in the posterior cortex, owing to the yellow colouration of the lens.)

The zig-zag ends may form a more or less regular complete wreath, behind and in front of the equator of the nucleus. It is not difficult to measure how relatively thin the flat opaque layers are in their middle and at their edges, in direct light as well as by transillumination. The opacification shows a tendency to progress axialward, especially in the posterior lens zone. At times I found this type complicated by fissures filled with fluid and with spoke formation, more often however, in fact almost *as a rule, in combination with lamellar separation* (see these). It may also be associated with coronar cataract. This flat wedge-shaped type of opacity is most common in the lower nasal zones of the lens.

Fig. 216. Flat sharply outlined wedge-shaped anterior peripheral opacities of the cortex.

Central to deep cortical area. Oc. 2, Obj. a2. The case of cataract with myopia, Fig. 226. This variety of cataract is not related to the type of cataracta complicata as found in degenerative myopia, it is an individually distinct type of senile lens clouding.

Note the upper angular and lower rounded cloud-like shape of the opacities. In the centre there are faint traces of spoke formation. Below this two (more lucid) dark stripes which remind one of the tears as seen in the separation of opaque fibres.

Axialward the lower opacity is bordered by a dense, white, sharply outlined zone. Toward the periphery it becomes more dense and concentric in position, in this manner surrounding the equator and joining the opacity in the posterior cortex. (The lamellar separation present in this case is not shown in the illustration.)

Figs. 217a and b. Flat, peripheral cortical opacity, with sharply outlined round and wedge-shaped border. W. age 65.

The opacity in this case, as is usual in this type, is found especially developed in a down and nasalward direction (a and b represent the lower nasal lens areas). There is lamellar separation in typic direction from below temporalward, upward to the inner nasal side. Axially the opacity ends in a wedge of more rounded shape. Note in Fig. "b" the concentric downward continuation of the stripe formation of this type of cataract.

The location of the cataract is in the middle and deep cortex. The very numerous lines of lamellar separation are in part anterior to the opacity. In the posterior cortex there is a similar type of cataract. Vis. both eyes = 1 without glass.

Figs. 218 to 224. Concentric lamellar separation. (Fissures formed between lamellae.)

This change is one of the most common types found in senile cataract. It is one of the typic changes. I found them at the time when I was still using the *Gullstrand* illuminating quadrant (slitlamp), in 1912, in various forms of senile cataract[81])[5]). Owing to their *folded-like appearance* I at first suspected the change to be due to a delicate folding, for the development of which the necessary substratum is lacking. The exact anatomic nature of this change is not fully understood.

To illustrate the real condition as well as possible, especially the effects of light and shadows, I have constructed plastic models to show just what can be seen. Photographs of these models are shown in Figs. 218 to 222, which are therefore presentations as they would be sketched direct from nature[81]). Figures 224 and 228 are however illustrations made directly from cases. One notes parallel fold-like lines (lines of separation), they are *usually* in the nasal lower lens zone arranged more or less in a steep manner, extending from below temporalward in an upward nasal direction. This must have some connection with the fact that in all of these cases there are flat wedge-shaped opacities in this same zone (Figs. 216 and 217).

Further observations show that at times the direction of the change is varied (Figs. 206 and 220). In Fig. 220 and 221 the separation turns abruptly into the fibre direction, so that we cannot speak of *concentric* separation, but are dealing with the separation of the radial lamellae of *Rabl*. In many cases I have found the opacities arranged in a manner like a spider-web, which is due to the fact that the lines of separation between the radial fibres are arranged and tensely drawn in a manner similar to the cross fibres between the radial fibres in the spider-web (Figs. 206, 222). At times I have seen a separation of the anterior nuclear surface, in which cases the fissures showed a similar spider-web like arrangement (Fig. 222).

That we are dealing with a lamellar separation I was able to prove by maceration experiments. The lamellae are separated by the absorption of fluid. These separated lamellae can be seen in *"optical section"* with the slitlamp bundle of light, just as we may observe in a like manner the fissures filled with fluid in the suture areas or between the lens fibres. Therefore, according to the direction of light, the lines of separation may at times appear to be superficial and again somewhat deeper in the lens. The sutures often show a separation, that is the edges of the seams are separated by fluid (the seams split), just as the lamellae (Figs. 206, 221, 222). Often the separation of the latter is *not complete,* so that diagonal and straight connections are found between the lamellae, comparable somewhat to the fissures in a traumatic cataract (Fig. 197), which are diagonally crossed by lens fibres (Fig. 224). The area of lamellar separation is usually in the cortex, central and deep. Where a separation is crossed by a seam the lamellar lines seem to sink in toward the seam and be attached to it (Figs. 220, 222).

In advanced cataract I found the nuclear surface showing lamellar separation. I also have frequently found diagonal separations inside of the fissures-filled with fluid and the spokes (compare remarks concerning Fig. 209).

As a rule cataract is present, usually flat wedge-shaped opacities in the periphery. Often opacities are found in front of and behind the lamellar lines of separation, or the latter are in the midst of an opaque area. In advanced and mature cataracts I have often found these lamellar separations.

Difficult to account for are the quite commonly found separation lines which are parallel and closely crowded in an otherwise transparent lens, *arranged in a straight line from one pupillary border to the other.* Interruptions or change of direction in the area of the sutures cannot be noted (Figs. 218, 223). Cases of this type cannot be sufficiently explained by accepting a separation of lamellae as the only causative factor in their genesis.

Fig. 218. Concentric lamellar separation of the anterior cortex (plastic representation). Prisca, age 68.

Right eye. Uniform cortical separation, from externally below extending up and inward, crossing a radial suture ridge in a smooth wave-like manner. A few circumscribed cortical opacities (not visible in the picture) are found. The direction of the separation is somewhat steeper above the radial ridge than below it. In the temporal direction beyond the lens centre the separation gradually decreases and is lost. Increased clouding of the cortex in this case as is usual, nasal and downward. Small circumscribed opacities are seen anterior and posterior to the separation area in the cortex. Behind the separated area in a definitely sagittal direction is seen the relief image of the nuclear surface. In this case it was easy to observe an apparent change in the location of the separation, by a variation in the angle of the axis of the illumination. (This also applies to the apparent location of the separations in Figs. 219 and 220.) In the left eye a somewhat more complicated separation or split (see Fig. 220).

Fig. 219. Concentric lamellar separation in the anterior cortex. (Plastic illustration.) B. age 71.

Right eye. Delicate "splittings" pass in a decided wavy manner over edges and depression of a radial fissure filled with fluid, *(w)* in the lower nasal cortex, in a direction from below and downward to up and inward.

Temporalward they become more delicate in nature and extend to the axis of the lens. An increased separation is shown in the lower section (at *sp*). On the nuclear surface there is a second, less developed design of separation (not shown). Nasal and downward is the area of greatest cortical opacification. The left eye also shows extensive separations. (Compare Fig. 221).

Fig. 220. Concentric lamellar separation of the anterior cortex. (Plastic illustration.) Prisca age 68.

Left eye. Coarse separations of complicated design in the cortex.

At "*W*" they extend, curving slightly, over the directly posterior suture ridge formation of the nuclear surface. At *R* and *R'* a typic division of the areas of separation, such as may often be seen in other cases.

The one group of separations takes a more radial direction following the fibres of the cortex (not of the nucleus!), at this place, the other group extends concentrically onward. At "*F'*" the direction assumed is also a radial one.

Fig. 221. Concentric lamellar separation of the anterior cortex. (Plastic illustration.) B. age 71.

Left eye. There is a double, in certain places a trifold arrangement of separation areas.

The nucleus and cortex show evidences of extensive pathologic changes, especially separations at the sutures. Fig. 221 shows the separation of the nuclear surface. A similar, more delicate designing of separations within the cortex is not shown.

Fig. 215—220. Tafel 24.

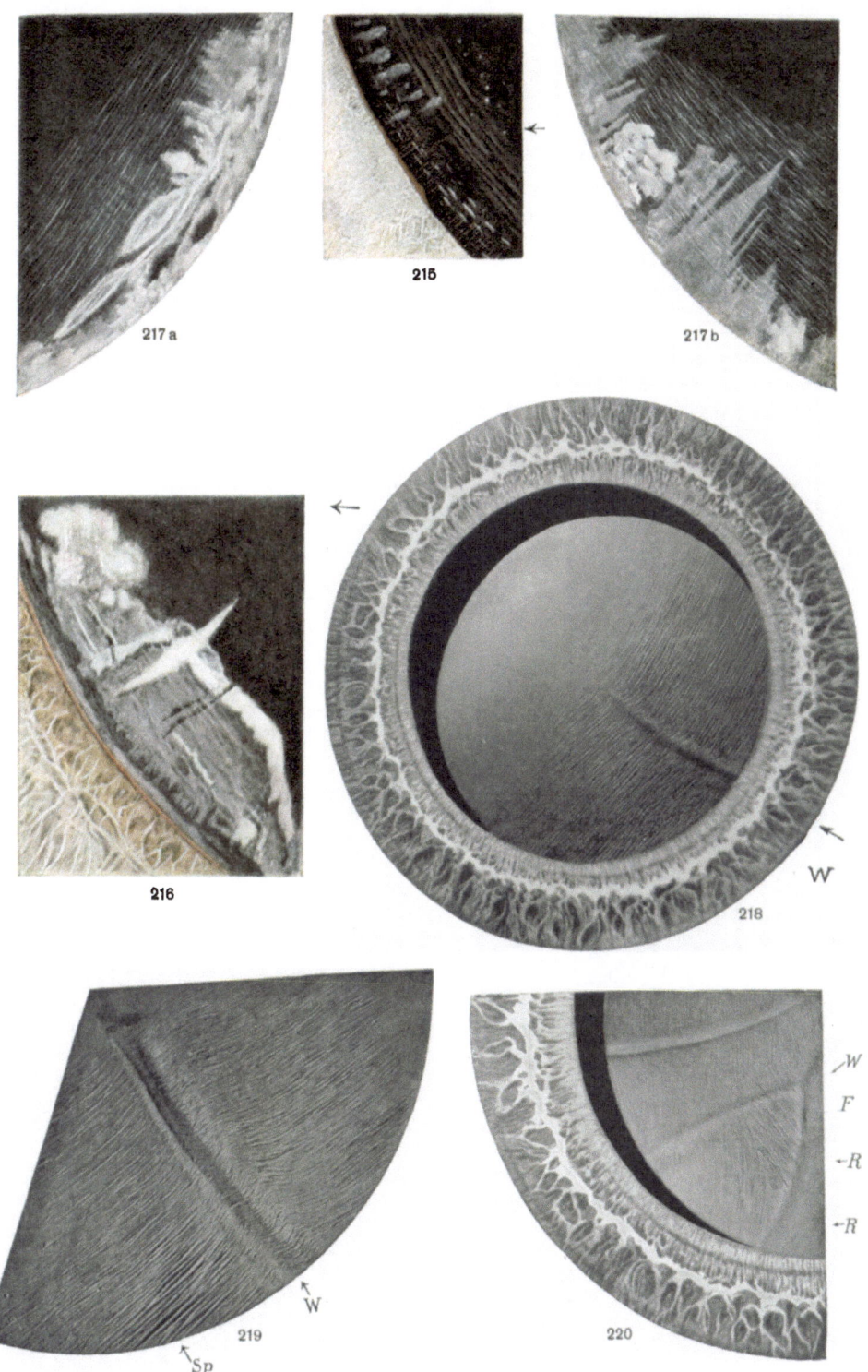

217a

215

217b

216

218

W

219

W

Sp

220

W

F

R

R

Verlag von Julius Springer, Berlin.

The separations are quite coarse. Nasalward they are covered by dense opacities *(T)*, which above the separations extend over two separated sutures, below they take the direction of the lens fibres.

The superficial "splits" of the cortex (not shown in the illustration) extend in a less steep manner and end abruptly at a separated cortical suture *(N)*. The nuclear surface at this point presents no suture, and the corresponding separation passes on uninterruptedly under the cortical suture.

Fig. 222. Concentric lamellar separation of the anterior cortex. (Plastic illustration.) B. age 68.

Right eye. The separations visible in both eyes are quite extensive, and in the right eye they present a double layer. Fig. 222 shows the axial nuclear area under high magnification (45 times). The lamellae seem forced apart, stretching across separated sutures. The separated surfaces of the lamellae enter into the suture fissures. Owing to insufficient separation one may note diagonal connections between the lamellae. The picture of a spider-net formation is brought about by the apparent separations connecting the sutures "*N*". In this case, as well as in several similar ones, I noted a structural relation existing between the lamellar separation and the knobs and granulations of the nuclear surface.

Fig. 223. Concentric lamellar separation of the anterior cortex. Very decided relief image of the anterior nuclear surface.

Case of Fig. 142, male age 73, left eye. Oc. 2, Obj. a2. Pupillary diameter 4—5 mm, light temporal. The lamellar separation "*L*" extends horizontally and diagonally from outward and below in an up and nasalward direction. In the area of the suture ridges of the nuclear surface "*N*", the lamellae become more delicate and seem to be depressed into the depth of the suture "*S*".

Note the contour of these pseudo-folds, illustrated true to nature, and the manner in which they seem to be mixed with one another, and again split up in certain places. These changes in this case are situated in the middle cortical layers, at least they are there visible if the direction of light and of observation are at an angle of 40° one to another.

Nuclear relief image and lamellar separation in this case are present in a corresponding manner in the *posterior cortex*. Peripherally (not illustrated) there are less dense flat zig-zag cataractous opacities of the type illustrated in Fig. 216.

Fig. 224. Concentric lamellar separation of the anterior cortex under high magnification.

Oc. 4, Obj. a3. Parts of the pseudo-folds (separations) are seen under a lineal magnification of 68 times. Note how individual lamellae split and show diagonal connections.

Fig. 225. Saucer-like posterior cataract. Mrs. M. G. age 60.

Oc. 2, Obj. a3. R.V. = $^{10}/_{200}$. L. aphakia following extraction.

One may observe in this picture a merging of two separate types of this form of cataract.

Section "*a*" shows the yellowish dense flat opacity, with vacuoles of small size in and on the surface. They are arranged mostly in rows, seldom in groups. As they are situated on a uniformly clouded yellowish *base*, they are visible in focal light by *transillumination,* that is in the light *reflected* by the opaque base. Therefore their dark outlines with a luminous centre.

In section "*b*" these vacuoles show a different optical behaviour. Here the opacification is not as advanced as in "*a*". (This area is comparatively clear when examined with the ophthalmoscope.) The spherical bodies, in focal light, inversely as shown in "*a*", now have dark centres and luminous borders, when seen in the same chosen angle of illumination. They are mostly large in size and in conjunction with their surroundings create the appearance of a sieve. This picture is characteristic of posterior saucer-like cataract. In the middle of this sieve the interstitial areas between the vacuoles are more dense and there is here a merging or transition with the dense type as shown in "*a*". In addition there is extensive dust and punctate coronar opacification, also of the deep peripheral cortex.

In cases of saucer-like posterior senile cataract there are old dense opacities in the deep equatorial cortex, which show a slower progression, compared to the saucer-like area. The latter change may be present in duplication, a similarly shaped opacity being found in a corresponding position in the anterior cortex. The *nucleus* is usually involved more or less and we may at the same time find all the other types of senile cataract complicating the picture.

Posterior saucer-like cataract is mostly situated in the subcapsular plane. Owing to the marked visual disturbances, it calls for early operative interference, often at a time when the anterior axial, cortical and nuclear areas are relatively clear, so that the pupil may still be quite black.

In differential diagnosis we must consider traumatic posterior rosette-shaped cataract and posterior cataracta complicata (compare Fig. 202, 268, 280).

At times one may note a distinct suture and fibre designing in this type of cataract. Regarding this form of cataract, compare Vogt[72]).

Fig. 226. Senile cataract complicating myopia.

Mrs. N. Elizabeth age 67, myopic since youth R. V. = $^1/_6$ myopia 10 D. L. V. motion.
Right eye. Cataract especially in periphery and posterior lens.
Left eye. Mature cataract—projection good, urine negative.

Uniform subcapsular vacuolar surface and sieve-like white subcapsular opaque network, the former visible in indirect, the latter by direct illumination.

The left eye shows a distinct, the right an indistinct iridescence of the anterior shagreen. Both lenses show a luminous anterior nuclear stripe (flat area showing as a line).

In the left eye there are white radiating lines and certain places show a typic lamellar separation. This left lens was extracted through an incision and conjunctival flap. The result was practically emmetropia. Vis = $^1/_3$ with + 2,00 axis 180. Faintly visible capsular remnants. The ophthalmoscope shows a large myopic crescent, a similar one is also visible in the other eye.

The right eye shows flat peripheral opacities anteriorly as well as posteriorly. The posterior area extends in the axial direction in the form of a rosette-shaped dense opacity (Fig. 226). The spokes of this rosette show a zig-zag irregular form. They are white, the interstices are less opaque, appear grayish and allow the passing of a faintly red fundus reflex. Axially this rosette is quite dense and thickened in a sagittal direction.

It is especially noticeable that the opacities show a distinct abrupt border and contrast sharply with the normal cortex. Anteriorly and posteriorly there is decided lamellar

Fig. 221—228.   Tafel 25.

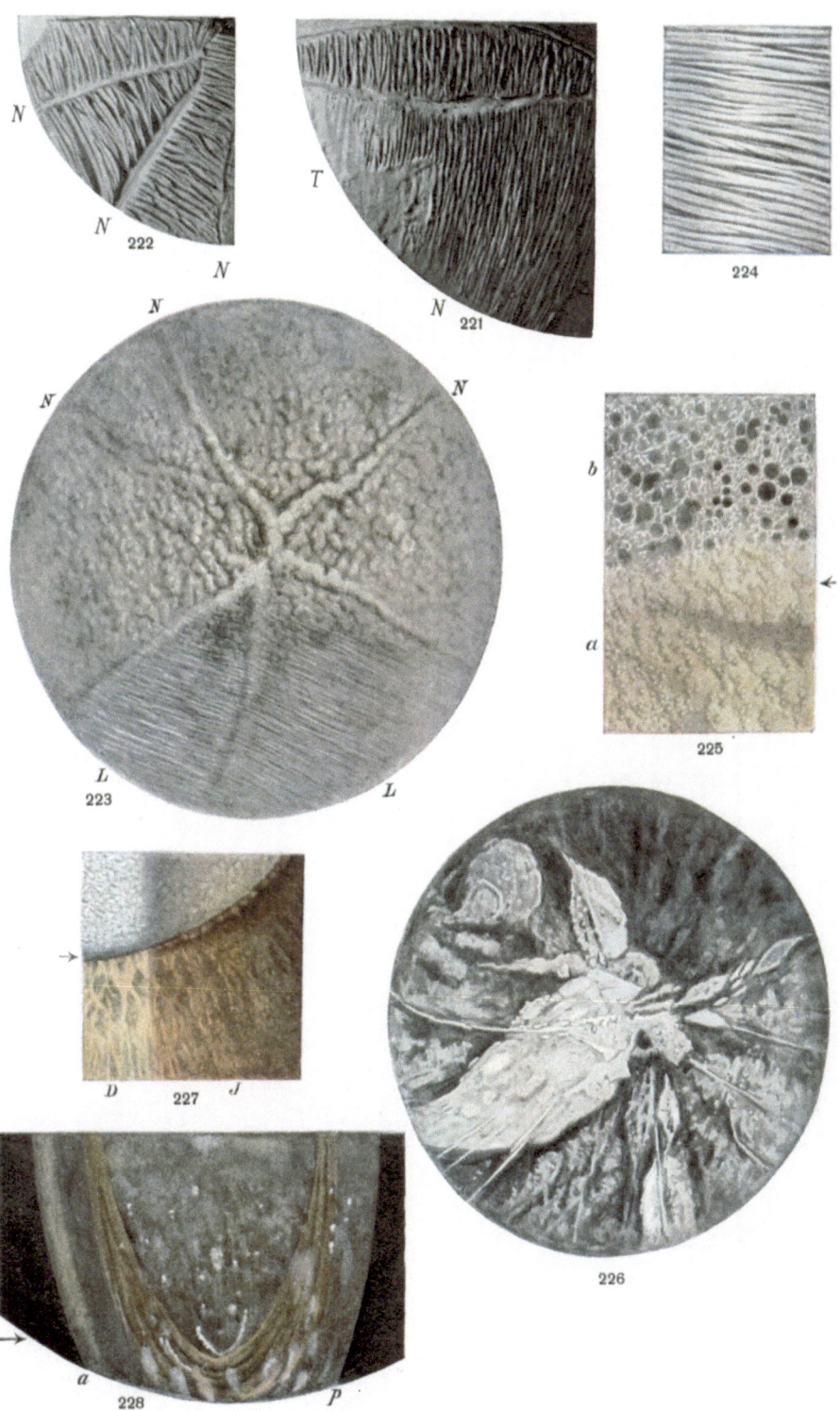

separation, a cataractous change found according to our observations in senile cataract, as well as in artificial maceration, but no doubt relatively rare in cataracta complicata. We are referring to the flat surface-like opacity as illustrated in Fig. 216.

That cataracta complicata occurs in degenerative myopia can be seen from the text describing Figs. 268 to 280. The combination of myopic cataract with the senile form is not rare.

Fig. 227 see also Fig. 279. Anterior subcapsular vacuolar layer.

This change is typic of the later stages of senile cataract. It occurs in cortical cataract (intumescence and mature), as well as in cataract of the nucleus (compare Vogt[21]). The observations were taken by transillumination, and especially with indirect lateral illumination, that is by direct or transillumination in the partial shadow at the *border of the bundle of light.*

One may see the dark, distinct outlines of droplets, small circular lines, often quite close to one another, without showing flattened sides. These myelin droplets (Morgagnian globules) as has been proved anatomically, are often in the epithelium, showing like sieve-like openings, in part under the epithelium, between the latter and the fibres and also in among the fibres. (I have seen similar globules occur as a postmortem change of the epithelium.) The diameter of these bodies is usually from 20 to 80 microns, one may observe larger and smaller forms (compare Fig. 279).

Often the deep cortex, at times including the nucleus, are white and densely opaque, while the superficial cortex may be quite transparent, with the exception of the subcapsular zone, which latter shows the vascular layer. The outer cortex is therefore free of opacities in these cases, in spite of which fact the cataract may be considered ripe for operation.

These vacuoles may be observed by utilizing the border of the bundle of light. They are as a rule invisible in *direct focal light.* However in more advanced cases they may show as white spots. (Compare Fig. 211, also the fissure vacuoles of Figs. 207 and 208.) They may be seen by direct focal illumination as a very faint grayish white subcapsular opaque network, a thin layer-like opacity, apparently in no direct connection with the deep areas of opacities, somewhat comparable to the saucer-like opacity of the posterior cortex.

By focussing the anterior graining (anterior reflecting area), the vacuoles are invisible, with the possible exception of those on the border of the layer. Iridescence of the anterior graining is usually, though not always present.

Fig. 227. Subcapsular vacuolar layer in mature senile cataract.

Mrs. B. age 66, right eye. Oc. 2, Obj. a2. (Regarding the atrophy of the pigment border of the iris of this case, compare chapter pertaining to this Fig. 302—307.)

To the left, direct *(D)*, to the right indirect illumination *(J)*. The vacuoles as such are visible only in indirect light. In the direct light one sees only a white spotted marble-like opacification. Compare also Fig. 208 and 211.

The anatomic examination of the freshly extracted lens has shown the dense cortex to be scatteringly filled throughout by these globules with fluid contents. Their specific gravity was greater than that of water, and they were micro-chemically identified as myelin droplets.

Fig. 228. Opacity of the peripheral concentric layers of the senile lens (Vogt[5]).

Oc. 2, Obj. a2. Miss M. K. age 55. Opacification of concentric layers is one of the common types of senile cataract formation. It can as a rule be diagnosed only by a full dilatation of the pupil. One sees in the deep peripheral cortex and in the equatorial area of the nucleus, large concentrically arranged flat saucer-like areas of yellowish-gray dust-like opacities, surrounding the nuclear equator. Only by accurate focussing with the bundle of light, can we discern the individual concentric onion-like layers. In addition to the dust-like opacities, one may often see larger dots. In the case of Fig. 228, which presents an optical meridional diagonal section through the lens (*a* = anterior capsule, *b* = posterior capsule), there are in addition a coronar opacification as well as dots and hooklets in the nuclear periphery. This type of lens change is diagnosable in its incipiency by the fact that the senile nuclear striping becomes more broad and luminous toward the equator. Clouding of the concentric peripheral layer type does not affect visual acuity until it has progressed sufficiently to involve the *axial cortex*. It can appear alone or in combination with other types of cataract.

Fig. 229. Nuclear cataract. (Schematic meridional section.)

The slitlamp has exposed the frequency of nuclear sclerosis. Observers still very recently have made the assertion that the nucleus as a rule is not involved in the development of cortical cataract. This is not true, as we have learned from observations with the Slitlamp.

The latter shows us for the first time an *increase in the internal reflection* of certain definite zones. The internal nuclear reflection is visible by slitlamp illumination, even through a comparatively clouded anterior cortex.

The increased internal nuclear reflection, which latter always inaugurates nuclear clouding, is first noticeable in the saucer shaped layers which bound the central interval anteriorly and posteriorly, that is in the *innermost area of the embryonic nucleus*.

During this period the central area is still discernible, owing to its darker colour (Fig. 229). In front of and behind it we see the embryonic sutures, the anterior a vertical, the posterior an inverted Y, which latter shows white on a less luminous background. Following this the *outer* bordering embryonic zone, that is the *outer saucer shaped (lamellar) zone* which *surrounds* the central nucleus becomes involved.

We now have the picture as is shown in Fig. 229, that is: a central zone of increased internal reflection, respectively a clouding, which is *separated from the senile nuclear area "N"* in all directions, by a more lucid interval *(J)*. This very characteristic picture may remain constant for weeks and months. The reflection of the senile nuclear area (strip), has also increased. Very gradually the heretofore clear interval *(J)* will develop an opacification, which now unites it with the senile nuclear area "N", leaving the internal *(R)* as the only separation between the clouded nucleus and the cortical strip "K".

On rare occasions I have seen nuclear cataract develop without a marked development of the clear interval *(J)*. In these cases it was immediately involved in the opacification of the other areas. The more solid consistency of the nucleus prevents it from changing into a soft mass, as compared to the cortex.

It is of especial interest to note that the development of nuclear cataract occurs simultaneously in anterior and posterior homologous areas.

For example, it never occurs that the anterior nuclear area becomes opaque in advance of the posterior, or vice versa. In the presence of an opaque cortex, the nucleus may remain clear for an exceptionally long time. Finally however it will participate in the clouding.

During this time it may occur that the central interval of the nuclear substance may split into an anterior and posterior half, as we have proved by experimental maceration (compare diss. of E. Meier[75]). This explains the rare occurrence of socalled "double lens" at the time of cataract extraction.

Though the clouding of the nucleus be so advanced that it may markedly involve the central visual acuity, the latter being at times reduced to 0,1 or more, we may still be able to *transilluminate* and *examine in direct focal light with the slit-lamp*, that is we may so to say "palpate" the form in! which the nuclear cataract is developing. At this stage the embryonic sutures are still visible as white lines.

Nuclear cataract is not alone a senile phenomenon, but may occur in cataracta complicata, especially in the rapidly developing types which complicate amotio retinae and iridocyclitis.

By means of our former methods only advanced cases of nuclear cataract were diagnosable, and only then if the balance of the lens remained comparatively clear.

The *Gullstrand* slitlamp has taught us the pathology of the nucleus in greater detail, also how to differentiate the cataractous from the ordinary sclerosed senile nucleus.

That the breaking down of the nucleus is not as extensive as that of the cortex, in which latter case a fluid emulsion may form, is explained by the difference in the consistency of the two substances.

The difference in the form of opacification is dependent on the variation in the aggregate masses. The disintegration of the nucleus occurs in the form of *dustlike* changes, the cortex is changed into coarse drop-like elements.

With the slitlamp we will in the future be able to sharply differentiate whether a nucleus be clear or opaque, normal or pathologic. The ordinary conception that a clear but *sclerosed* nucleus is void of vitality and therefore to be accepted as a foreign body within the lens substance, is thereby proven erroneous. I would consider this *sclerosis* as a *process of conservation*. It protects from disintegration! Opacification of sclerosed areas is never as intense as that of the cortex.

Fig. 230. Anterior pyramidal capsular cataract of unknown origin.

Oc. 2, Obj. a2. Left eye, pupil $2^1/_2$ mm. Miss. B. R. age 56. At 5 years of age she was supposed to have suffered an attack of keratitis. Both cornea show circumscribed areas of slight opacification. No evidences of perforation.

The broad based opacity of the axial area shows a blunt knoblike, densely white, in part pigmented projection, extending forward into the aqueous. Note the suggestion of radial and concentric striping (lamellation?).

The surrounding capsule shows scattered pigment. The anterior graining, iridescent in part, is separated from the opacity by a dark, less luminous area or layer. I find this present in all cases of so-called capsular cataract and consider it characteristic of the latter (compare text of Fig. 184). To the left and below there is an additional small opacity. Vision $= ^1/_6$.

There is a second intensely white opacity, not visible in the illustration, in the area of and somewhat behind the anterior senile nuclear strip, connected with the capsular opacity by a *constricted area* so that the full opacity is shaped like an hour-glass.

There is no doubt that this constriction developed during the growth of the lens, from the pressure exerted by clear lens substance. This constriction may possibly in other cases lead to a complete separation. (I have a record of several cases of this nature.)

The slitlamp picture of this case tells us that the cataract developed at a time when the fibres of the present anterior senile nuclear surfaces were still adjacent to the epithelium. The other eye has a similar small capsular cataract *confined* to the *capsular area.*

Fig. 231a and b. Cataract complicating myotonic dystrophy Dr. F. B. age 60, right eye. (J. G. Greenfield, J. Hoffmann, Fleischer[142]*) and others.)*[*]

Oc. 2, Obj. a2. This opacity is of the type of coronar cataract. There is however this difference that in the coronar cataract of this case the opacities are only peripheral, while the axial cortex is permeated by dustlike and short lineal opacities. The appearance of some of the punctate opacities in certain areas is very striking, they show as vivid greenish and reddish cholesterin crystals formation. The dustlike opacities are so densely situated and numerous that both the anterior and posterior cortices seem veiled by them, producing an extraordinary picture of ocular pathology.

The senile nuclear areas or zones are luminous, but indistinct. The posterior embryonic zones also show a vivid reflection. The senile nuclear relief image is well defined. The opacities in the anterior and posterior cortex show a preference for involvement of the *deep areas.* They are also larger in this zone. Here again the deep equatorial cortex shows the greater involvement. There are large opaque dots measuring as much as 0,1 mm in diameter, also larger and course coronar opacities in this area. Nasalward and below there are less distinctly seen gray lancetshaped stripes. Fig. "*b*". Posterior below and also nasalward there is a dense spoke. (To the left in Fig. "*b*".)

$$RV = 0,5 \atop LV = 0,5 \Big\} \text{concave, } 1,50 \, D.$$

Fig. "*a*" shows a surface area of the lower quadrant in the right anterior cortex, magnified 24 times.

Fig. "*b*" in contrast to this shows an optical section of the same lens area. The white stripe "*K*" is the anterior capsule. One may see in its medial area a few punctate opacities shining through it. To the left of this stripe one sees the optical section "*Co*" through the cortex, with its punctate dust-like, and gray flat striped opacities.

This cortical lens strip ends abruptly at the nuclear border. (The nuclear relief image does not extend sufficiently toward the periphery to show this.) The nucleus "*N*" is clear, behind it extending into the cortex one sees the before mentioned spoke.

In spite of its peculiar characteristics I do not definitely wish to designate this type of cataract as pathognomonic of myotonic dystrophy.

The sister of the patient has similar lens opacities and muscular disturbances. I found opacities of the coronar type in a large number of members of this same family. The father was supposed to have had myotonic dystrophy, probably an older brother, as well as a younger sister of the patient.

---

[*] Regarding the history of this cataract family, compare Vogt, Zeitschrift f. Aughk. Vol. 40, book 3, page 133. The medical adviser of Herr F. B. Professor Egger, Director of the General Policlinic Basle, confirmed our diagnosis of myotonia atrophica.

Fig. 229—235.   Tafel 26.

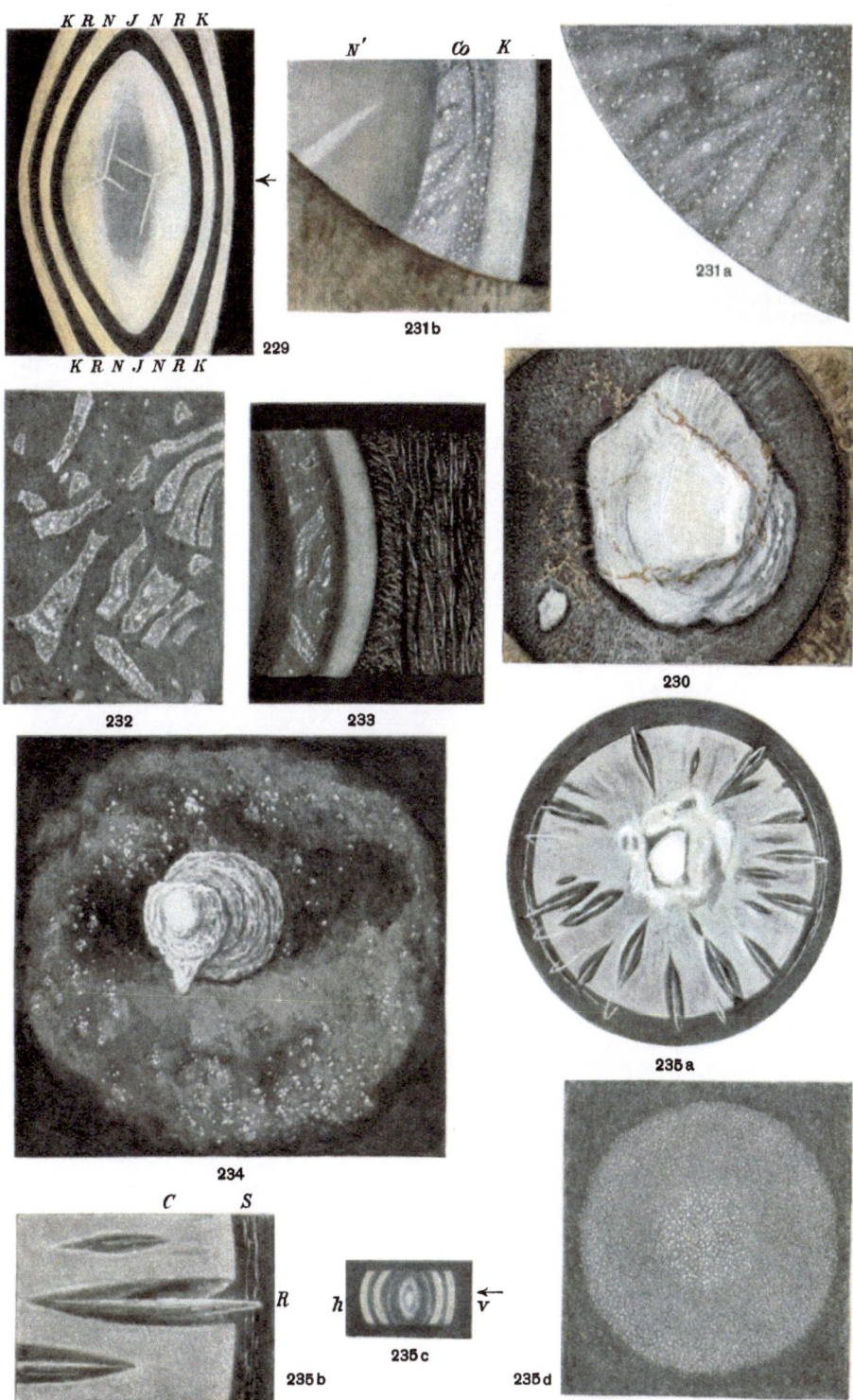

KRNJNRK

N'   Co   K

231a

229

KRNJNRK

231b

232   233

230

234

235a

C   S

R

h   v

235b   235c   235d

I have since observed another case of this type .of cataract in a 52 year old woman Mrs. R., a member of a large family-group with cataracts, certain members of which suffered with "muscular atrophy", and simultaneously presented psychic disturbances. (Two brothers and a sister of the mother.)

On the mother's side: the great grand-father, grand-father, a grand-uncle, mother, uncle, and an aunt were all operated for cataract in their fifth decade of 'life.

The lenses of this Mrs. R. show the following characteristics: The nuclear relief image is distinct. The cortex is permeated by an enormous number of dust-like opacities, in part iridescent. Peripheral concentric lamellar clouding, a few of the coronar type. Posteriorly, below a few spokes. Posterior axial zone shows a $^1/_2$ mm porous cloud (such as is seen in cataracta complicata), and an iridescence of the posterior axial reflecting zone. The vitreous was normal.

Vision R—5,00 = 0,25.

Vision L—2,00 = 0,8.

In this case the similarity of the type of cataract with that of Fig. 231 induced me to search for instances of muscular dystrophy in the members of this family-group.

Figs. 232 and 233. A rare type of juvenile cataract ("Ribbonlike juvenile cataract"). J. J. age 22.

Both lenses show sharply outlined gray stripes, irregularly placed, of peculiar shape, in the area of the posterior surface of the senile nucleus (Fig. 233). (At times their direction seems to correspond to that of the lens fibres.) These stripes are very thin in the sagittal direction, and present peculiar, concavely bent-in ends. In the stripes there are many white spotted and punctate opacities. A few of these are found outside of the stripe.

These ribbon-like opacities are found in both eyes in the same corresponding layer and are bounded posteriorly by clear cortex. Evidently they originated at a like period in both eyes. The patient came under our observation on account of a right bulbar contusion. (Compare punctate deposits in vitreous, Fig. 233.)

An inexperienced observer may have ascribed the lens changes of the right eye to the effect of the contusion, had he not found the corresponding changes in the left eye.

Fig. 233 shows the opacification in a diagonal sagittal optical section, for the purpose of localization. Fig. 232 presents the opacities extending in a plane surface.

Fig. 234. Posterior capsular cataract, probably congenital. Mrs. K. R. age 25, right eye.

This amblyopic eye was operated on for concomitant strabismus during her 9th year. At that time an opacity was diagnosed at the posterior pole.

The snow-white peak-shaped opacity rests with its broad base on the inner polar surface of the posterior capsule, extending pyramidally forward into the cortex, to the area of the embryonic nucleus. The posterior embryonic suture is not visible. The base of the pyramid measures 1 mm. The sagittal length is approximately 1,5 mm.

In the anterior direction the peak tapers in a flat angular manner, so that a pyramid is simulated. The base of the pyramid is surrounded by an opaque halo (see Fig. 234), which latter I have found present in other similar cases. This translucent, very thin, opaque zone is in the area of the posterior capsule. It is sharply outlined peripherally.

Visus: R = $^1/_4$, L = 1. The other eye shows no peculiarities. In this and in other similar cases the slitlamp shows a layer-like structure of the zones which are parallel to the posterior lens surface.

At times I saw the cataract extend forward in the form of a pointed "runner" which reached the centre of the lens, giving a spindle shape to the cataract. This extension proves that the beginning of the change may be traced to an early embryonal period.

Figs. 235a and b. Zonular (lamellar or perinuclear) cataract, with a circum-scribed opacity of the anterior capsule. Miss S. K. age 5 has reduced visual acuity, but no ocular disease.

In addition to the bilateral typic lamellar opacification there is an anterior capsular cataract.* This is shown by the white spot in the middle of the picture. By focussing the anterior shagreen one may observe that it is surrounded by a dark halo (compare text to Figs. 184 and 230).

Surrounding this opacity there is a distinct iridescence of the shagreen. The lamellar opacity merges axially with this capsular opacity. *It therefore extends anter-iorly in the axial zone toward the capsule. Posteriorly there is a similar axial extension toward the capsule.* In spite of this opacity it can be seen that there is a clear zone between the cataract and the posterior capsule. Peripheral to the anterior capsule in the equatorial zone the lamellar cataract is separated from the capsule by a clear zone. The zonular cataract shows in this as well as in other cases, as a thin scale of dense opacification surrounding a nearly clear nucleus. Around this sharply outlined opaque scale, there is a faint irregular veiled layer of minute droplets, too small for measurement, which contains the socalled "riders".

The latter are equatorially situated, opaque hooklets with pointed ends, which straddle the opaque lamella. *It must be emphasized that the hooklet in this case always sits on a fissure,* the clear spaces, which are the result of separation of opaque fibres (see Fig. "a" and "b").

These fissures are found uniformly separated in the equator of the before de-scribed opaque scale or lamellar zone, and have been described anatomically by Peters. (Sutures are not visible).

Some fissures are found outside of the equatorial areas, they also have over them a parallel opaque stripe, the "riders" (Fig. 235 b).

How may we explain the relation of riders and fissures filled with fluid? I consider the riders, after the beginning of fissure formation, as being composed of newly developed fibres extruding from the opaque nucleus. Regarding the extrusion of clear substances from opaque, see the text to Fig. 277. (All cases do not neces-sarily show fissures under each rider.)

In this case, as well as in similar types of cataract I have often found the lens flattened antero-posteriorly. This corresponds with the well known conclusion that in perinuclear (zonular or lamellar) cataract the lens suffers in its development.

Figs. 235c and d. Cataracta centralis pulverulenta. Probably congenital in origin. Apprentice H. B. age 18.

Similar conditions in both eyes. "C" presents an optical sagittal section through the centre of the lens, under low magnification.

---

* As is well known, the opacity in capsular cataract is situated below the capsule in the changed hypertrophic epithelium. The capsule in itself is always transparent.

$v =$ anterior, $h =$ posterior, d shows the lens opacity from the anterior surface. The equatorial diameter of the whole opacity is 2,2 mm; the central more dense opacity measures 0,8 mm. The balance of the lens in both eyes is clear. Embryonic sutures not visible.

The opacity is composed of dustlike and punctate dots. As can be seen in Fig. c, the central opaque nucleus is more rounded in the sagittal direction, compared to the less dense layer or shell peripheral to it. One can speak of a shell as there is a more dense border demarkating it from the clear peripheral cortex (Fig. c). Fig. d shows the equator increased in density because of the duplication due to the curving of the opaque layer.

During the course of one year's observation the cataract had not changed in the least. Vision under mydriasis $= 0,8$, otherwise $= 0,5$. With the ophthalmoscope this opacity is noticeable as a faint shadow. If one accepts a subcapsular origin for the opacity its development would correspond to the first embryonic months. The patient shows no abnormalities otherwise.

Figs. 236 to 259. Anterior axial embryonic cataract.

The type of cataract I have described by this name[52]) occurs in 25% of all healthy persons. Its location in relation to the central interval is shown by Fig. 236 (low magnification). On account of the axial position of this type of cataract, pupillary dilatation is not necessary to facilitate its observation. Investigation of a large series of individuals has proved to me that this type of lens change has no lowering effect on central visual acuity. Its time of origin probably corresponds to the period just following the separation of the lens vesicle from the wall of the secondary optic vesicle. (Fig. 236 is a schematic sketch of the sagittal lens section magnified 10 times. Figs. 256 to 259 show the cataracts magnified 37 times.)

The other illustrations (lineal magnification 24 times, Oc. 2, Obj. a2) show the anterior embryonic cataract to be composed of scattered snow-white foci forming a group within the borders of the central interval, or rather inside of the lamellar zone which forms the anterior border of this interval, which therefore is in the area of the vertical suture of the upright Y. Less frequently the opacities are situated *within this central interval.* (This is the case in Figs. 254 and 255.)

An intensive white colour especially characterizes this type of opacity. With the exception of calcification and in capsular opacities, I found equally white changes only in early punctate and hookform opacification of the equatorial cortex and of the nucleus. In older persons this white colour may appear yellowish in tint, owing to the colouration of the lens substance anterior to it.

The individual foci may be absolutely separated from one another (as in Figs. 239 to 244, 252 to 255), or they may in places be confluent or connected by a veil-like clouding (as in Figs. 237, 238, 249, 256 to 259).

Often, though not always, these foci are situated in one and the same concentric surface (in an embryonic nuclear surface), or there may be additional smaller foci, slightly anterior or posterior. These areas of foci are separated but little in the sagittal direction. They never extend to the area of the posterior embryonic suture. At times the small foci form a group, which latter is then situated axially directly anterior to the central interval, or possibly somewhat inside of the latter. (For instance in Fig. 238 and 246.)

The individual areas are composed of intensely white spots and dots, connected by a veil-like opacity. (The diameter of these foci varies from 0,05 to 0,5 mm.) Often the dots show a colourless glitter, at times I saw coloured crystals, such as are called cholesterin. In other cases certain dots resembled droplets. They may form dense groups or be arranged in a scattering manner. Very characteristic is a *white halo* surrounding these white areas. This halo gradually merges into the surroundings (Figs. 237 to 250, 256 to 259). It is not always equally distinct, and at times only visible by intense illumination.

The arrangement of the individual zones near the anterior Y suture is easily observed (Figs. 237, 244, 245, 250). It can occur that the whole Y figure appears as an opacity (Figs. 237, 256). The fibres as they leave the sutures or seams may participate in the opacification (Fig. 256). Therefore the opacity may appear shaped like a hand, or feather, and it may be branched (Figs. 245, 247, 256).

In some cases the ends of the Y suture may be joined by a series or line of punctate dots (Fig. 245).

Cases of this kind may be of interest in efforts to explain the origin of this type of cataract. I have found this anterior embryonic cataract in children, whom it was just possible to examine with the slitlamp on account of their extreme youth. In these they show the same characteristics as in individuals 70 years of age, and older. *We are therefore, according to my observations, dealing with an absolutely stationary type of cataract.*

Through the kindness of *Prof. Hedinger*, Director of the Institute of Anatomy and Pathology, Basle, I came into possession of two bulbi of a little girl, two months old, which showed this type of cataract in its definitely clear characteristic form. I show these cataracts as they appeared after the cornea was dissected off of the fresh bulbi (Figs. 256, 257, magnification 37 times).

At a certain distance from the superficial surface one can see a luminous lamellar separation (Figs. 256, 257 at D), which was seen before the corneas were removed. It probably corresponds to the area of separation I have described. The embryonic cataract was situated somewhat nearer to the anterior capsule than it is found in older individuals, which is to be expected in view of the character of the development of the lens. The cataract in this child was in the corresponding place and of the same type as is found in older persons. Its appearance in a child two months of age makes it practically certain that this embryonic cataract presents a congenital lens change. The embryonic cataracts here illustrated were observed, partly in the eyes of persons employed in this institution, partly in clinic and poli-clinic patients. Its common occurrence makes it quite easy to readily obtain any number of cases.

I will therefore not go into further detail describing the cases here illustrated and will refer the reader regarding the probable genesis and a complete description of this type of cataract to the "Zeitschrift f. Augenhk." 1919, Vol. 41, Page 125.

Figs. 260 to 266. Cataracts of the anterior and posterior embryonic suture.

M. van der Scheer[83]) has proved that this form of cataract is typic of *Mongolian Idiocy.* It has been also described by *Pearce, Rankine* and *Ormond*[84]) and *B. Leeper*[86]).

It is situated in the embryonic suture area, as I have been able to prove. That this type of cataract also occurs in mentally and physically normal persons is proven by the illustrations. At times the anterior, again the posterior, or both embryonic suture areas may be involved. The sagittal distance separating the opacity from the central interval is greater than in the types of anterior axial embryonic cataract

---

* In the literature this cataract is described as *Cataracta stellata* (Greeff).

Fig. **236—261.**   Tafel **27.**

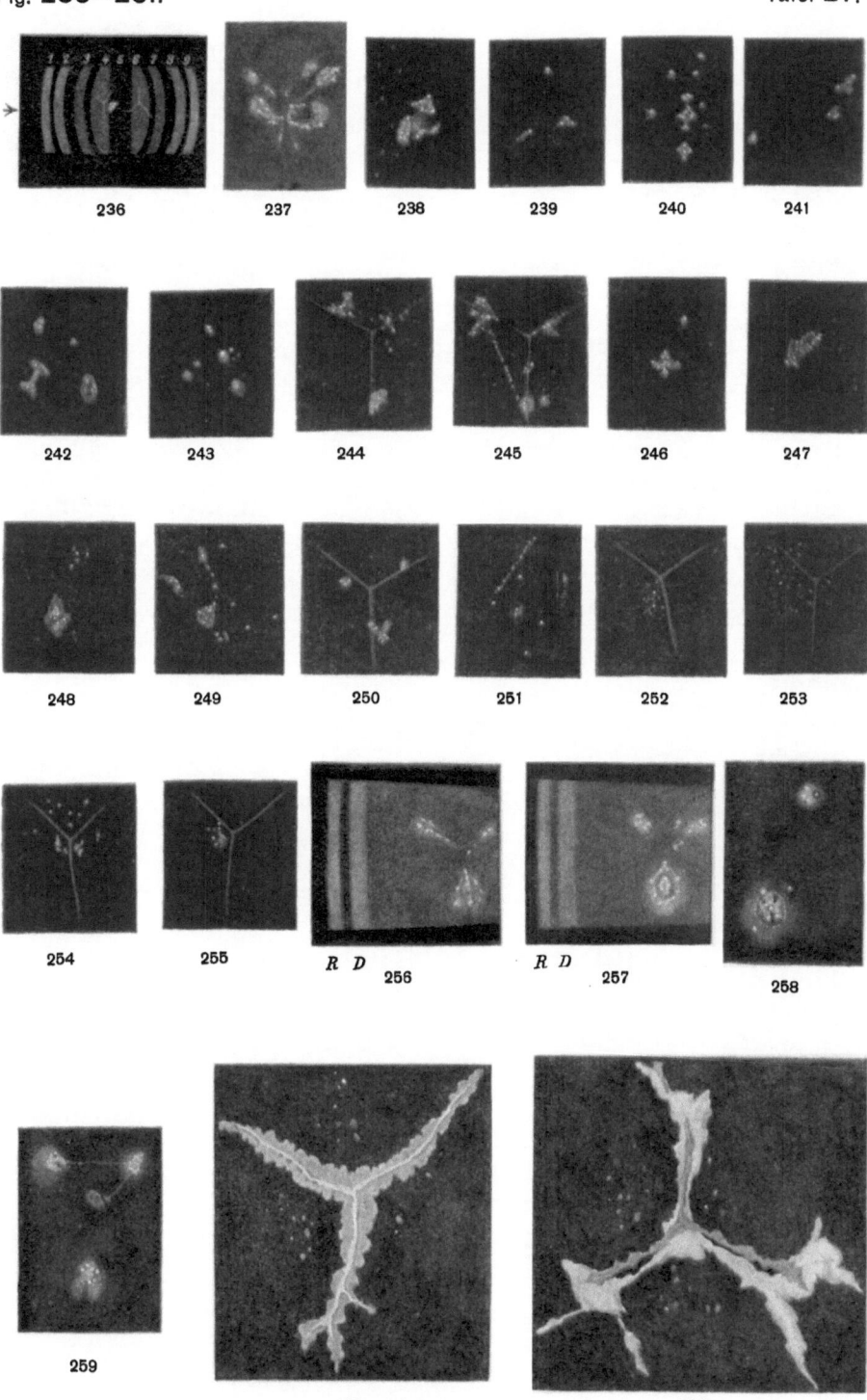

236   237   238   239   240   241

242   243   244   245   246   247

248   249   250   251   252   253

254   255   *R D* 256   *R D* 257   258

259   260   261

shown in Figs. 236 to 259, and the areas of extension of the opaque branching is increased. The individual opaque rays at times measure 1 mm or more. This proves that the probable time of origin of this type of cataract corresponds to quite a later period than the time of genesis of the more delicate changes of the first described form. If one accepts the fact that the embryonic suture cataract (Figs. 260 to 263), develops in the subcapsular area, its time of origin would correspond to the end of embryonic life or to the time just following birth.

Another point of differentiation is the fact that the anterior axial embryonic cataract *occurs only in the area of the anterior embryonic suture*, or in the anterior part of the central interval, while the embryonic suture cataract is found both anteriorly and posteriorly in similar degrees of development. That the clinical picture of the two types is quite different can be seen from the illustrations, and for this reason confusion is impossible. Further observations may teach us if there be transitional types of the two.

Figs. 260 to 263. Bilateral anterior and posterior embryonic suture cataract, F. E. age 25.

Oc. 2, Obj. a2. (The opaque arms of Fig. 260 are about 1 mm in length.) Both lenses show numerous opaque stripes in the anterior and posterior embryonic sutures. (Fig. 260, to the right, anterior, 261, to the right, posterior, 262, to the left, anterior, 263, to the left, posterior.)

The opacities develop a certain width and thickness and show a definite double layer, characterized by a different density and colouration.

The layer nearest the central interval is more translucent and shows a blue to bluish-green tint. The external layer is thicker and of greater density, and shows a white to yellowish colour by focal inspection. Its arms corresponding to their more excentric position are larger and show an increased branching. In other words the bluish translucent opacity corresponds to a younger lens area, a stage of lessened suture length compared to that of the white, more dense opaque layer. All of this is shown by Figs. 260 to 263. In the anterior embryonic area the white opaque layer is *in front* of the bluish one, in the posterior *behind it.* (This similarity of position of these two like changes within homologous anterior and posterior areas, suggests a separate time of origin for these types of differently coloured opacities.) Note the constrictions which show as a kind of a segmentation in the cataract of the anterior suture. Near the sutures one sees opaque spots, which in part follow the fibre direction. The eyes of this patient are normal otherwise, and he is physically and mentally sound.

Fig. 264. Anterior embryonic suture cataract, R. S. age 49, left eye.

Oc. 2, Obj. a2. For the anterior embryonic suture there is substituted a the bluish-green opacification, varying in density. The borders show a moss-like, delicately branched configuration. Near these there are individual isolated pale cloudings, similar in appearance. The posterior embryonic suture, also shown in the illustration, is intact. This cataract differs from the ones shown by Figs. 260 to 263 by virtue of the described moss-like type of opacification. The patient was normal in other respects.

Figs. 265 and 266. Posterior embryonic suture cataract, Miss H. von A. age 16.

Oc. 2, Obj. a2. Directly behind the central interval (Fig. 266 left eye) there is a white moss to grill-like branched opacity, which in its form reproduces the axial area of the posterior embryonic suture. Fig. 265 shows the opacity of the right eye. It is present only in a rudimentary form. There are two spots of opacification outside of the posterior embryonic suture, they are however of the same type as shown in Fig. 266. In the centre of the posterior embryonic suture there are a few punctate opaque dots.

Fig. 267. Spotted cataract of the anterior embryonic nuclear zone, F. K. age 65.

Oc. 2, Obj. a3. At a certain distance behind the senile nuclear surface, near the anterior embryonic suture, on a surface presenting an anterior convexity, there are very thin opaque blue-green dots of irregular form, visible in focal light. Posterior lens zone clear. Peripherally a few coronar opacities of the deep cortex and nucleus. This cataract is probably an early type of cataract of the embryonic nucleus.

With these illustrations have been shown only a small part of the many types of juvenile and embryonic cataract presented to us by the slitlamp. However they are the most common types of these varied forms of cataract.

Figs. 268 to 280. Cataracta complicata. (The cataract due to exogenous noxious substances, in contrast to senile cataract.)

Regarding the iridescence of the posterior lens area in cataracta complicata compare Fig. 194—196. The characteristic feature of cataracta complicata (compare Vogt[72]), within its strict limitation, that is the type of cataract which complicates disease of the retina and the uveal tract in amotio retinae, retinitis pigmentosa, iridochoroiditis, chronic chorioretinitis, chronic glaucoma, degenerative myopia etc., is its incipiency in the area of the posterior lens sutures. (Less frequently in the anterior.) The opacity is often *rosette-shaped*, as it extends in the direction of the sutures, and shows an irregular porous, tuff or pumice-like structure[72]). The more dense areas of the opacity appear like a mass of crumbs or seeds.

This porous structure is due to the development of vacuoles. Especially in the axial direction and in the area of the sutures, the opacity extends *diffusely forward into the cortex*, in the form of very irregular crumbly white opacities, surrounded by a thin veil-like nebula.

The porous structure and the tendency to extend forward into the cortex, are especially important points of differentiation between it and senile cataract, especially the saucer-shaped posterior cortical type. In the latter there is no axial thickening, that is the anterior extension into the cortex, which is so characteristic of cataracta complicata. Often the most advanced opacification in cataracta complicata is not situated exactly at the posterior pole but somewhat on one side, at times in the periphery, in the areas of the suture ends. The axial opacities may present a somewhat circular ringform arrangement.

The anatomical structure of the lens is not brought out or emphasized by its extension, as we see it in the progress of senile cataract.

Very rarely do we see it extend in definite concentric zones or in the direction of fibres. However the sutures are somewhat favoured, therefore the rosette-shape characterizing this type of cataract. Often one may see early in its development, two or more step-like flat layers, sharply demarkated as such, but indefinitely outlined, of which the posterior layer is the one of greater density. The crumb-like, intensely white areas in this cataract may be beginning calcareous deposits.

A very vivid iridescense of the reflecting zone, in the area of the posterior pole, often also peripheral, has always been found wherever the capsular area was lacking in that degree of involvement, which made its observation still possible.

Fig. 262—270.

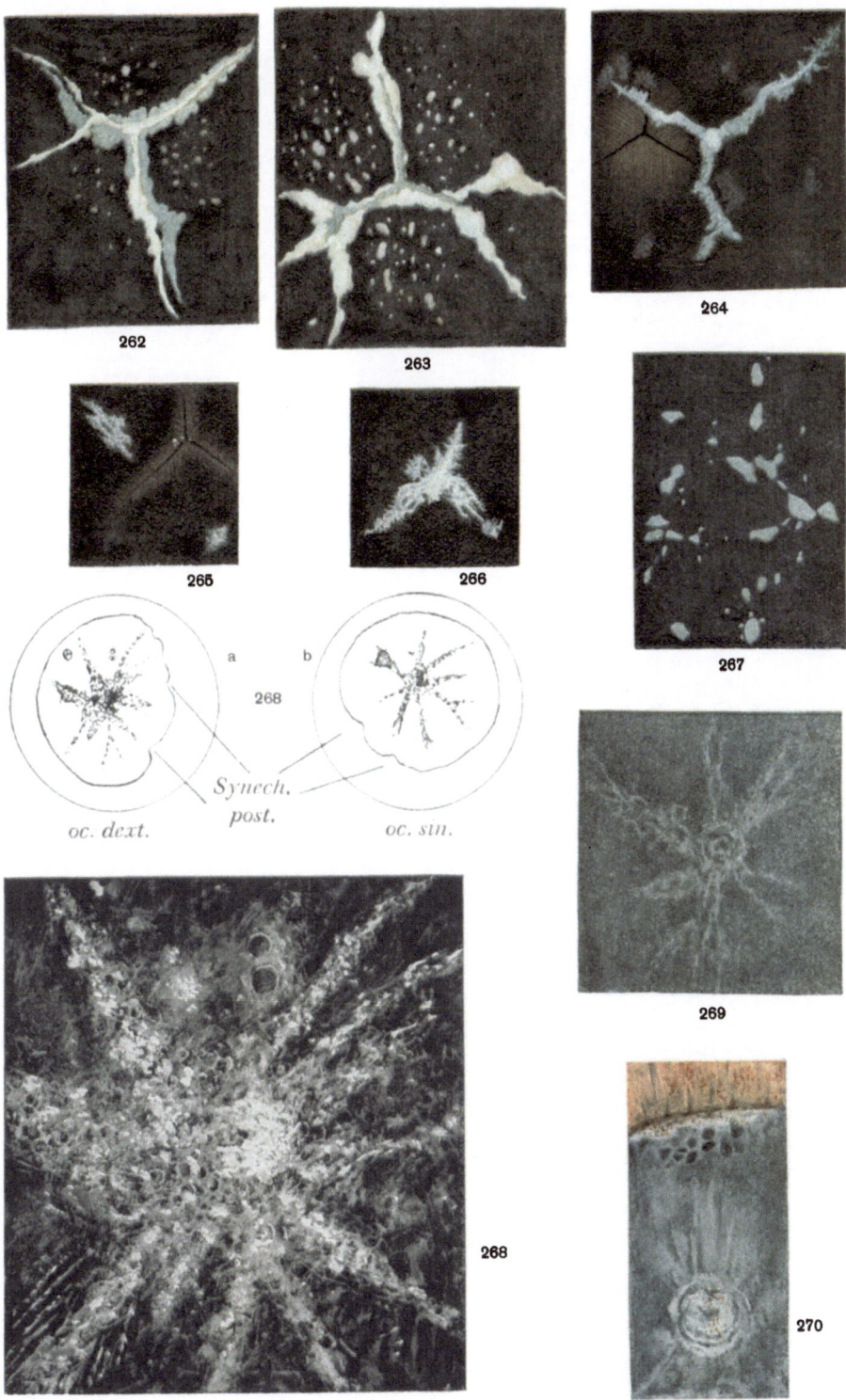

262

263

264

265

266

267

a   268   b

Synech.
post.

oc. dext.   oc. sin.

268

269

270

Verlag von Julius Springer, Berlin.

Anteriorly in the lens one may find at an early stage, often later, a more or less rosette-shaped irregular opaque zone, favouring the suture areas in its choice of involvement. This opacity in its incipiency is very faintly luminous. In the anterior, as well as the posterior areas it is noticeable in early cases, that the thin nebular cloudings are not alone subcapsular, but extend somewhat deeper into the cortex.

Large vacuoles are seen in later stages.

Rapidly progressive forms as we see them in amotio retinae and iridocyclitis often combine themselves with nuclear cataract. This change however occurs only in the later stages. It is self-evident that cataracta complicata may at times be combined with senile cataract. Old stationary forms of traumatic cataract (see Fig. 200) may show the type of the cataracta complicata as it has been just described. Regarding incipient cataract complicating heterochromia, compare the chapter pertaining to the iris Figs. 304 to 306.

Fig. 268a, b, c. Posterior cataracta complicata in chronic iridocyclitis. Mrs. A. K, age 51.

The chronic inflammation has existed in both eyes for the past twenty years. Fig. *a* and *b* show the cataracts in both eyes by transillumination with the ophthalmoscope. There are present posterior synechia of the iris, peripheral coronar and superficial anterior cortical, as well as posterior rosette-shaped opacities. In the posterior cortex, the cataract is denser axially and parallaxially, and it shows a club-shaped, dense spoke formation.

Fig. 268c shows the axial area of the posterior rosette opacity under a 24 times lineal magnification. Porous structure and vacuole formations are suggested. The dense white, often seed or crumb-like zones represent necrotic, if not in part, calcareous fibre areas. The cataract extends in the direction of the sutures and *at the same time diffusely forward into the cortex*, in the axial as well as in the suture areas. The cloudy porous pumice-like areas at no place show a sharply demarkated border. Upward there are two large flattened vacuoles, to the left and downward there is an area of opacification in the fibre direction. In this as well as in other cases of this type of cataract, there are isolated foci of dense areas of extension, near the rosette-shaped rays (Fig. *a* and *b*), which latter show a club-shaped thickening in that respective direction.

In this case the axial involvement has extended to the posterior nuclear zone in the form of increased opacification. The very vivid iridescence of the posterior reflecting zone is not shown in the illustration. It was visible axially as well as peripherally.

Fig. 269. Incipient anterior rosette-form cataracta complicata. Mrs. K. age 55.

I had treated her during a period of two years for chronic iridocyclitis (on the basis of Tbc.). For the past two years the eyes were free of precipitates and irritation. There are remnants of posterior synechia. In the area of the axial anterior cortex there are by focal illumination faintly luminous, by transillumination with the ophthalmoscope invisible, grayish opacities, which show a distinct rosette-form in the left eye.

The centre of the rosette is situated slightly upward of the lens pole, and consists of two indistinct concentric ring-formed opacities. The radial stripes are in part very superficial, in part they extend into the deep cortical layers. Here and there one may see minute droplets. Oc. 2, Obj. a2.

Fig. 270. Anterior incipient cataracta complicata.

Right eye of case 269. Oc. 2, Obj. a2. Above the middle of the anterior cortical surface there are three concentric ring-formed opacities in the superficial cortical layers. Anterior of the middle of this opacity there are a few brownish red pigment areas on the capsule. Extending up and down from the ring-formed opacity there are a few delicate opaque stripes. Above, near the pupillary border, which latter is covered by pigmented exudate remnants, there is a network of faintly luminous opacities which surround dark areas (vacuoles of the superficial cortex). As the eyes are quiescent it can be presumed that the cataracts in this case, as well as in the case of Fig. 269 are stationary in character.

Fig. 271 a, b, c. Posterior cataracta complicata.

a = View of the posterior rosette-formed cataract by transillumination with the ophthalmoscope.

b = Part of the lower inner opaque area magnified 24 times, with focal slitlamp illumination.

Bo. Jos. age 29. Rifle shot $4\,^1/_2$ years ago. The bullet is supposed to have hit the outer orbital border, squeezed the bulb from the side and *luxated* it anterior to the lids. (Scar in skin and bone in the area of the outer orbit.) Ophthalmoscopically in place of the temporal retina there is a greenish white avascular tissue mass, which latter is separated from the uninvolved fundus zone by areas showing choroideal changes. Yellowish atrophy of the papilla, dustlike vitreous opacities. Vision = 0.

The rosette shape of the posterior cataract is due to the porous opaque masses which extend in a characteristic manner, especially in the direction of the sutures. In the areas of the suture ends, the opacities are in part of greater density than axially. Here they are *most dense* not subcapsular, but in the direction of the senile nuclear surface.

Vivid iridescence of the posterior reflecting zone, also in the periphery. Regarding the anterior opacity see Fig. 272. The internal lens reflection is increased compared to that of the other eye. It is absolute and quite marked in the area of the posterior senile nuclear surface.

The posterior embryonic nuclear surface shows an abnormal form. Instead of presenting a concavity anteriorly it presents it *posteriorly*, causing this surface to merge with the senile nuclear surface in the periphery (Fig. 271 c).

The other eye presents the same peculiar abnormality. In how far the bulbar concussion in itself contributed to the cataract formation, cannot be estimated at this time.

Fig. 272. Anterior cataracta complicata.

Case of Fig. 271. In addition to the extensive posterior rosette-shaped cataract, there are delicate anterior nebulae, axially—in part subcapsular and partly in the adjoining cortex. Oc. 2, Obj. a2. By focal illumination one may see three round white rings of opacification like strung pearls and in their arrangement following a suture. Modified grayish fibre bundles radiate from the latter to both sides. There is iridescence of the anterior reflecting zone of the lens. I presume that the ring-shaped opacities are remnants of former fluid deposits (vacuoles), which after having been flattened, had their contents absorbed. (Compare also text to Fig. 276.)

Fig. **271—277.** Tafel **29.**

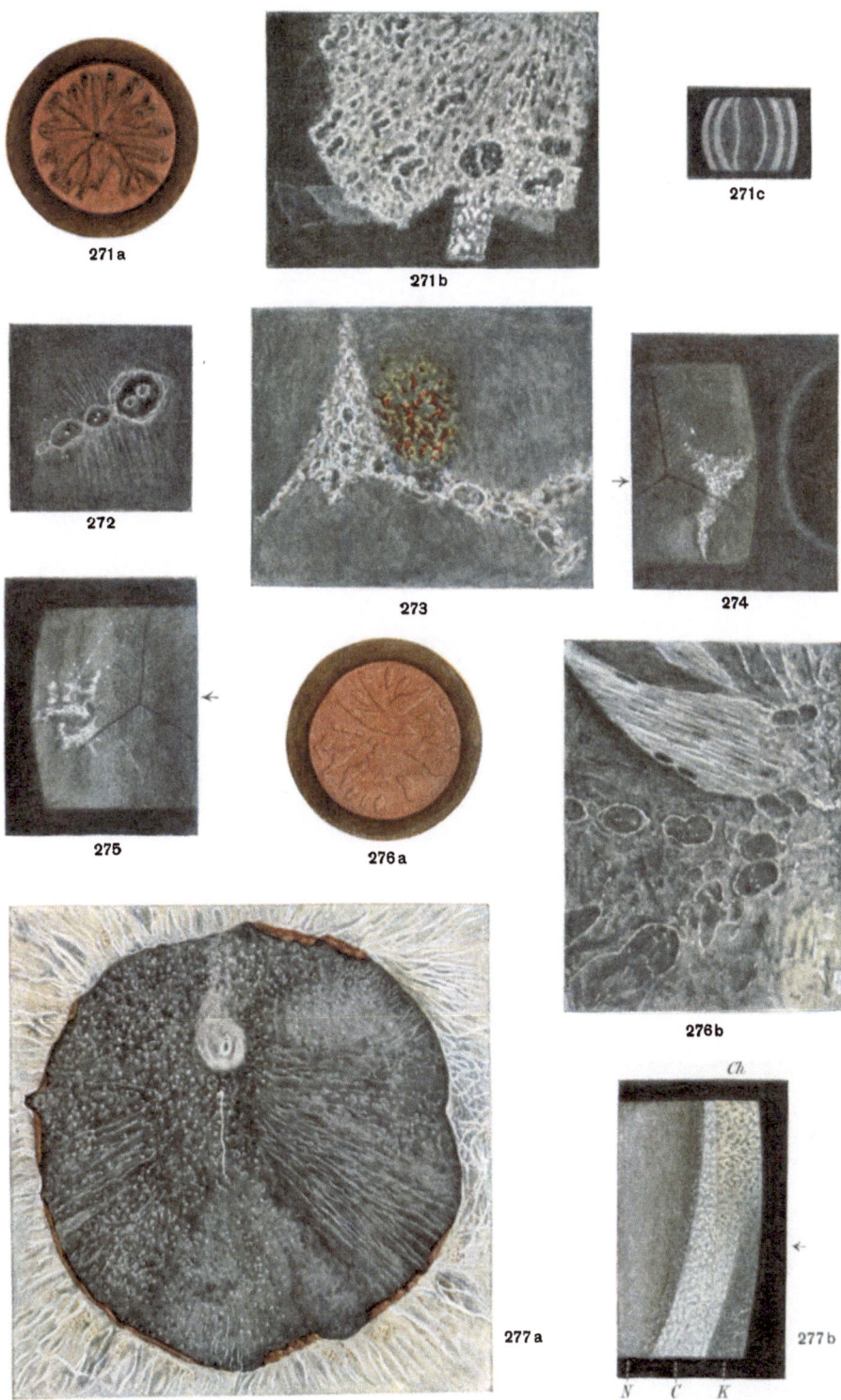

271a

271b

271c

272

273

274

275

276a

276b

277a

277b

Ch

N C K

Verlag von Julius Springer, Berlin.

Fig. 273. Incipient posterior cataracta complicata in retinitis pigmentosa.

Mr. Sch. age 39. Oc. 2, Obj. a 2. Right eye. Only the axial area of the cataract is shown. Marked iridescence of the posterior reflecting zone. The cataract is triangular axially and has three tip-like extensions corresponding to the posterior sutures.

It is composed of porous white dense wasp-nest like opacities extending into the cortex. The three radiations show branching, the lower nasal larger ray can in part be seen in the illustration. The whole cortex is somewhat cloudy, partly filled with more dense opacities. The supporting structure of the vitreous shows delicate punctate deposits and there are preretinal irregular lineal reflections. With the red-free light one may note cystic degeneration of the macula and many typic peripheral pigmentary changes. The other eye presents a similar condition. Vision each $= {}^6/_8$ without lenses. Concentric contraction of the visual fields.

Figs. 274 and 275. Incipient cataracta complicata in degenerative myopia, with retinal and choroideal hemorrhages of the macular area.

Mrs. R. age 60. Lowering of the visual acuity on account of circumscribed hemorrhages, for the past year many ragged and dustlike opacities in a fluid vitreous. Fig. 274 right eye, Fig. 275 left eye.

Oc. 2. Obj. a 2. The opacities at the posterior pole are not visible by transillumination with the ophthalmoscope. They are grayish white, cloudy, in part porous in nature, varied in density, in places just like a thin veil.

They are situated exactly at the posterior pole. In the illustration, the bundle of light is directed axialward from the temporal side, while the direction of observation is diagonal to it, so that the embryonic surface seems laterally displaced. The opacity at the right has an apparent vertical extension of ${}^{20}/_{24}{}^{ths}$ of a mm and horizontally one of ${}^{10}/_{24}{}^{ths}$ of a mm. The one at the left measures ${}^{11}/_{24}{}^{ths}$ of a mm horizontally, although delicate lineal opacities extend somewhat beyond.

To the right there is also a delicate curved line (compare Fig. 166—178). It is somewhat broader than usual and its vertical extension measures ${}^{40}/_{24}{}^{ths}$ mm. A vivid iridescence of the posterior axial reflecting zone is not shown in the illustration.

Fig. 276 a and b. Incipient, rapidly progressive, cataracta complicata in a recent case of amotio retinae.

Miss L. W. age 23. Myopic since youth, suddenly developed cloudy vision in her left eye, four weeks ago. There is an extensive area of amotio retinae in the lower fundus. Refraction at the papilla — 2,00 D. With the ophthalmoscope one may observe a layer of vacuoles (Fig. a), which, as can be seen by the slit-lamp, are situated under the left posterior capsule. Two weeks ago the *subcapsular opacity was of slight density and extension.* Today it shows a decided progression.

The layers of fluid are very flat in places, and therefore show colours of dispersion (interference), with yellowish and yellowish-red shades predominating. Especially on the *borders* of these surfaces there are white porous opacities of varied density. They are just adjacent to the capsule and extend somewhat into the subcapsular area. (Fig. b shows the vacuoles of Fig. a, nasalward of the posterior pole by slitlamp illumination.)

It can be easily seen that *every vacuole is surrounded by a circular opacity.* (Compare ring-formed opacities in Figs. 269 and 270.) The cataract therefore first develops at the *border* of the vacuoles. The central parts of the fluid layers also show a slight clouding. The large horn-shaped opacity in the upper part of the illustration, within it, has grayish white opaque stripes in the direction of the fibres. The vitreous presents the changes in the supporting structure and the reddish-brown deposits which are characteristic of amotio retinae. (Compare chapter pertaining to the vitreous.)

Fig. 277a and b. Old cataracta complicata after bulbar contusion.

(An addition to the method of determining the depth of lens opacities, compare the introduction of the chapter pertaining to the lens, and Fig. 212.) A. K. age 46, was hit with a stick in the left eye ten years ago. Vision $= \frac{1}{200}$. Many old tears in the sphincter of the iris. Fundus indistinct on account of lens opacification.

Fig. 277a (Oc. 2, Obj. a2) shows the ruptured sphincter of the 4 mm wide, sluggishly reacting pupil. Iris structure in the areas of the rupture is atrophic. Pupil border adherent to the lens capsule above and temporalward. In the upper pupillary area there is a dense white round opacity measuring $\frac{1}{2}$ mm. The shagreen (graining) is absent over the centre of this opacity, while it can be observed in its periphery. A sharp focussing of the (narrowed) focal bundle of light onto the anterior capsule (see text page 9 and the beginning of the chapter pertaining to the lens), shows the following:

Directly under the capsule there is but little opacification visible. To the right and upward there is only a faint opaque network containing vacuoles. Aside from this there are only a few dots and stripes in the lens area under the capsule.

The anterior shagreen is iridescent. The real opacity is a thin flat layer and it is situated at some distance from the capsule. This distance is somewhat increased toward the periphery, and corresponds to the area or surface of separation. (Compare Fig. 100b.) There is therefore inserted between the capsule and the opaque area, a layer of clear lens substance, thin axially and becoming uniformly thicker toward the periphery. As the injury occurred ten years ago, could the clear lens substance not be composed of newly developed lens fibres which have interposed themselves between the capsule and the opacity which latter at that time was in a directly subcapsular position? (Compare to this my experimental evidence: the development of cataract by ultra-red rays.)

I am also in possession of a second similar case. There is therefore a form of contusion cataract, which develops faint, thin, flat stationary opacities, which spread under the surface of the whole of the lens capsule, and which later on, on account of the growth of the lens, may be wholly or in part forcibly separated from the capsule. (Compare also the text to Fig. 314.)

The cataract in this case is of an uncommon type. It consists of a cortical layer of dense flakes of uniform size. The individual flakes have a diameter of 40 to 80 microns, some less. As far as can be observed the posterior cortex is involved in a similar manner and it shows a vivid iridescence. There is a pigment deposit on the lower posterior lens capsule. The senile nuclear surface appears indistinct. As can be observed in the illustration, the intensity of light varies in places, in the anterior cortical opacity. In addition there is opacification in the fibre direction, especially to the right and downward. Here the floccules are decidedly smaller. Ophthalmoscopically this is hardly discernible. It can be presumed that the cataract is now stationary in form. In regards to the technic of illumination this case is very instructive. If one would use the slitlamp as a simple form of illumination, (in which manner a certain observer recently endeavoured to solve the question of the genesis of senile cataract[87]), without understanding the principle of the use of the lamp), one would come to the erroneous conclusion that the opacification has a directly subcapsular location. The opaque surface is uniformly smooth and under full mydriasis there are no factors present, by means of which we may expect to find or be. able to diagnose any other but a directly subcapsular location. In this case however there happen to be the very few dense subcapsular stripes which may be of aid. How do we therefore determine the location of the capsule?

As I have explained on page 8 and as is demonstrated by Fig. 277b (Oc. 2, Obj. a2) we at first *sharply focus* a reflecting area of the anterior capsular surface, *by exact regulation of the narrowed bundle of light*. This is attained when the lateral border lines *a c* and *b d* (Fig. a) of the anterior capsule stripe are in sharp focus. The angle between the line of observation and that of illumination is quite large, about 60°, and we so direct the patient's eye, that the direction of vision will bisect the radius of the lens surface under observation.

Fig. 277b shows an optical meridional diagonal section through the lower pupillary area of the first eye. The pupil is dilated to 6 mm, and the source of light is from the temporal side.

One may see to the right the anterior capsular area, which in its upper part only, shows the shagreen "*Ch*". The white stripe "*C*" to the left is the surface of the cataract and the presence of a clear cortical area between it and the capsule, and its increasing thickness peripherally, is very evident and shows plastically by binocular observation.

At "*N*" the clear nucleus.

Also note that of the capsular area "*K*" one only sees the right edge distinctly, while the left edge is seen very poorly with its luminous opacity.

It however shows an immediate definite contrast if we change the focus to the reflecting zone "*Ch*"! (By narrowing the slit a more definite survey of the arrangement of concentric layers may be made than is shown in Fig. 277b.)

Figs. 278 and 279. Cataract in siderosis.

J. R. age 26 sustained a perforation of the cornea, iris and lens of the right eye by a splinter of iron three years ago. Today the eye is free of irritation. It has light projection. Iris dilates incompletely under atropine. The relief or design of the iris stroma is indistinct. The tissues of the iris are yellowish-green to brown. The other iris is grayish-blue with a few pigment areas. The wound of perforation of the iris is temporal and downward (see Fig. 278), opposite that of the cornea (Fig. 27). Cortical cataract with fissures filled with fluid (Fig. 278 above), showing red rust spots, situated in the superficial layers.

These foci are indistinct in outline and show an even distribution in the area of the pupillary border. At certain places it can be seen that these foci are arranged in their grouping in correspondence with the fissures which are filled with fluid. They originate near the sutures, the parts of the lens which present a "locus minoris resistentiae" to the absorption of pathologic fluids. Therefore, in all probability, the regular circular arrangement of the foci of rust deposit. A rust deposit is also found at the anterior pole where the sutures by virtue of their junction predispose to an early formation of fissures filled with fluid.

Fig. 279 shows a red rust deposit under high magnification (Oc. 2, Obj. a2). One may see by focal light that the cataract is composed of a subcapsular surface of vacuoles, which are in part covered by these rust spots. I could not find any of the latter deposits in the capsule. It is well known that these rust foci are due to the deposit of iron compounds in the proliferated capsular epithelium (compare *E. v. Hippel*[88]).

Fig. 280. Cholesterin crystals in advanced cataracta complicata.

Oc. 2, Obj. a2. J. W..age 68, right eye. Myopic since youth. Vision $R = \frac{1}{200}$ coronar cataract, cataracta complicata posterior and in part anterior. Nuclear cataract with extensive development of cholesterin crystals.

The crystals are flat rhomboid, in part triangular. In many places they seem arranged concentric with the fibre direction, especially so in the equatorial- area of the nucleus. With the slitlamp the picture reminds of vivid luminous christmas-tree decorations.

In cataracta complicata as well as in senile cataract, especially if hypermature, this development of relatively large cholesterin platelets is quite common. Small coloured crystals are also quite common. I have found them in normal lenses in youth. Their chemical nature has not been determined.

Fig. 281. Rare type of saucer-shaped juvenile posterior cataract probably due to tetany.

A. J. Locksmith age 30, physically somewhat below par (Oc. 2, Obj. a2). A part of the opacity of one eye (nasalward) is shown. The opacity apparently is subcapsular. The patient for years has been subjected to intermittent trophic disturbances in the growth of all finger nails. There is also intermittent alopecia. Other symptoms of tetany were not found. Urine is normal.

The reduction in visual acuity has been noted for the past three months. Both eyes show extensive saucer-shaped posterior cataracts, axially of increased density. In the axial opaque area there is a dense vacuole formation, peripherally the opacity is more translucent, and fibre striping is discernable. $LV = \frac{6}{60}$.

A senile nuclear strip (surface) is not visible anteriorly nor posteriorly. The embryonic nucleus seems flattened. Both lenses are extremely thin in their anteroposterior diameter. The latter is 3 mm measured according to the method as described on page 4.

I extracted the cataract of one eye. *It showed that the whole lens substance was sclerosed.* A black pupil was the immediate result. Measurement of the hard yellow nucleus showed 7 by 3 mm. Fundus normal. Vis $= \frac{6}{24}$ with $+ 12,00$ D.

The cause of this peculiar type of cataract is unknown. A family tendency as far as could be investigated was not proven. It must be presumed that the primary change is an *early sclerosis.* The patient insists that he saw well up to three months ago. One may associate the trophic disturbances in the finger-nails with an inhibition in the lens growth, early sclerosis of the whole lens, and a destruction of newly developed lens fibres. It is self-evident that the sclerosed lens substance is not subject to this destruction.

CATARACT ARTIFICIALLY PRODUCED IN RABBITS BY EXPOSURE TO SHORT-WAVE ULTRA-RED AND LONG-WAVE RED LIGHT (BEYOND WAVE LENGTHS OF 670 μμ).

The experimental production of cataract by means of this exposure to light is based on our former investigations regarding the passage of ultra-red light through the eye media[89], [90]. These have proved, contrary to the formerly accepted belief, *that the ultra-red (short wave length) of our artificial sources of light reached the lens and retina in a greater quantity than the volumn of visible light.*

Fig. **278—287**.

Tafel 30.

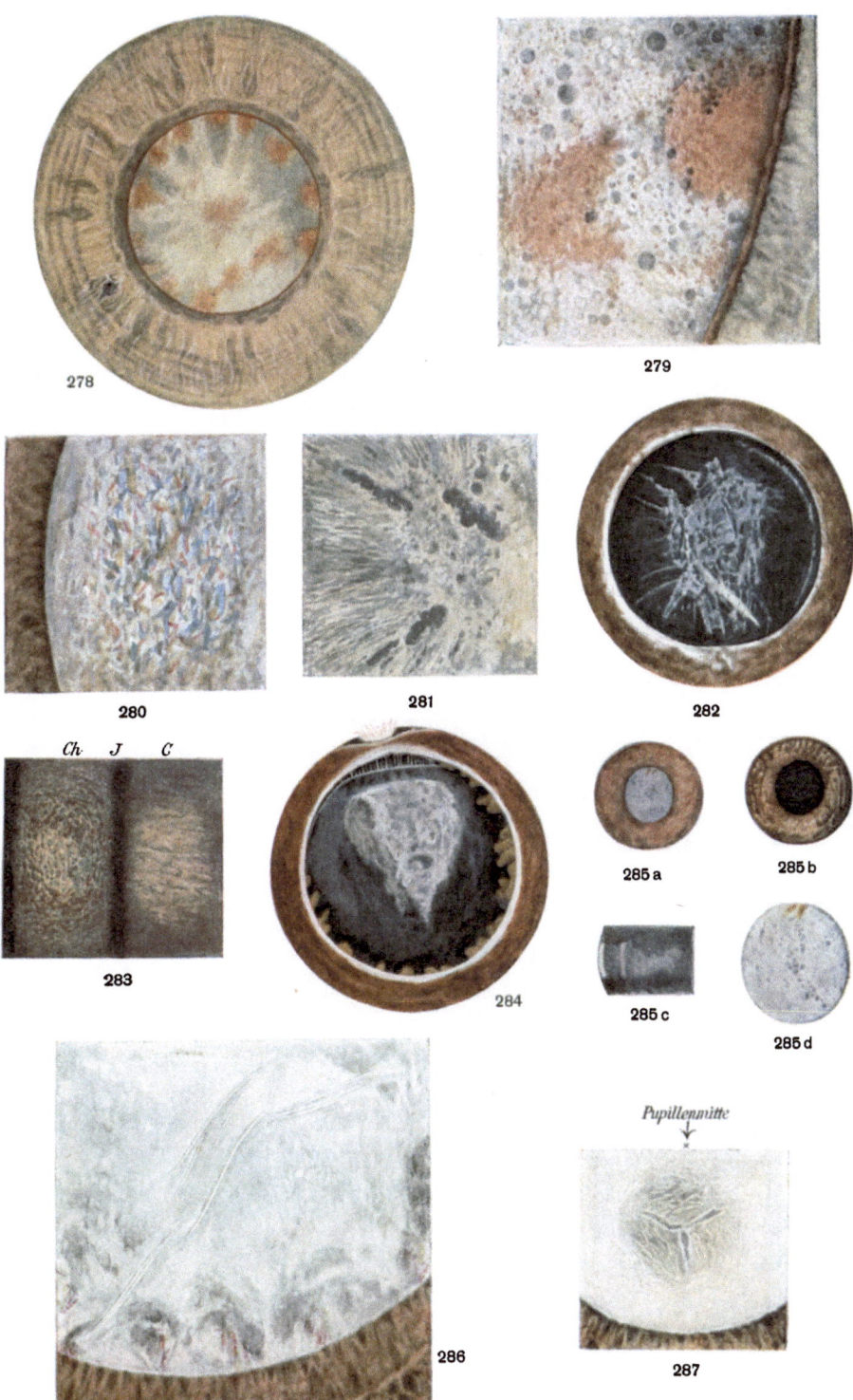

278

279

280

281

282

Ch   J   C

283

284

285 a

285 b

285 c

285 d

286

Pupillenmitte

287

Verlag von Julius Springer, Berlin.

By using several water filters, a few centimeters in thickness, these short-wave ultra-red rays which pass the eye media, were in part absorbed. We have proved by former experiments the damaging effects of short-wave red on the iris. By a half hour exposure of rabbits' eyes to the ultra red of an arc-lamp we have produced an irritation of the iris lasting for hours, in fact for a duration of days.

By an improvement in our methods it was recently possible for us to expose eyes to this light for *much longer periods*. We now eliminate the formerly troublesome factor, the heating of the filters, by a method of drainage, whereby the fluid is constantly changing. This allows us to expose the eyes for any desired length of time. The light of an arc-lamp of 30 ampères is passed through a filter chamber of clear mica. This filter chamber has two compartments, each having a lumen of about 1 cm. One compartment contains flowing cold water, the other a flowing solution of iodine and potassium iodide. (Iodi. pur., Kalii iodati āā 50,0, Aqua font 100,0.)

The latter solution allows only long wave red to pass (areas of absorption between 670 and 700 $\mu\mu$).

Concentration onto the eye is accomplished by means of rock-salt lenses. The red that still passes is for the purpose of directing the exposure to the eye.

The crater of the direct current arc-lamp is at about 6 to 7 cm distance from the filters. One the back of the latter is placed the lens of natural sodium chloride. The iodine filter absorbs not alone all visible light, with the exception of the outer red, but also all ultra-violet, and together with the flowing water the long wave ultra-red. The exposure of the eye is therefore to the short-wave red to which is added some red of a greater wave-length[92] [93]).

These very lengthy, time-consuming exposures were in greater part accomplished by Dr. U. Lüssi, assistant at the clinic.

The thermic action of the concentrated filtrate on the skin can hardly be noticed. In spite of this we have produced a total lens opacification by a three hour exposure in a black rabbit.

We have been able to produce cataracts by this method in any desired number of rabbits.

In the cases in which we have created partial cataracts, they were as a rule present in the anterior as well as in the *posterior* cortex. The posterior cataract was situated opposite the anterior one, in correspondence with the direction of the rays. The cortex seemed less resistant to the influence of the rays than the nucleus. The opacities were directly under the capsule as well as in the deep cortex.

In young growing animals especially, observing them during the course of weeks and months, I could see this subcapsular opacity advance further into the lens, being forced away from the capsule by the interposition of newly developed lens substance. This experimental observation is important for the purpose of better explaining the genesis of certain types of cataract, for instance that of the lamellar form. Compare the analogous observations made on injuries to the lens-capsule (*Leber* 1880), and on Roentgen-cataract (*E. v. Hippel* 1905 and 1907).

In albinos we were not able to produce cataracts in as short a length of time. Cataract was more easily produced in older rabbits, compared to the time found necessary in young animals. Anatomical investigations of the cataracts are still incomplete.

By these methods it was for the first time possible to produce cataract experimentally, with the aid of the rays combined as described. The latter are important elements of natural and of artificial light.

We may now safely infer that the socalled glass-blowers cataract is due to the effects of *ultra-red* and *ultra-red + red rays*. This form of lens sclerosis is not due to the effects of ultra-violet rays as had been taken for granted.

In spite of the many experiments in that direction it was not possible to produce cataract with ultra-violet light. It would however be an error to apply our experiments in adopting conclusions as to the genesis of senile cataract. For this other additional factors must be considered. The conclusion that cataract occurs at earlier ages in India and other warm climates, if this be so proven, can gain *no support* from our experiments. Our investigations do not exclude the possibility that other than ultra-red and red, for instance other visible rays, may also produce cataract. This has however not been proved. (Regarding the production of cataract by means of pure sunlight compare *Czerni* 1867. Regarding that produced an arclight filtered through water see *Widmark* 1889—1901, and especially *Herzog* 1903.) Regarding the *pigment* and other *proliferating changes* we have produced by ultra-red + red rays in the iris, see this chapter.

Fig. 282. Anterior cortical cataract produced by exposure to ultra-red and red rays in a full grown brown rabbit.

Observation several weeks after exposure. Low magnification. The axial cortical zones were especially exposed.

The posterior cortical cataract is as dense as the one in the anterior cortex. One may see at times sharply circumscribed, slightly curved linear opacities, which resemble *spokes*. The nucleus is clear, it is more resistant to the exposure than the cortex.

Regarding the pigment changes in the iris see the chapter pertaining to the iris.

Fig. 283. Iridescence of the anterior axial shagreen. Anterior and posterior cortical cataract produced by exposure to ultra-red.

Full grown brown rabbit, nine days after exposure (Oc. 2, Obj. a2).

Light from the left. To the left the shagreen *"Ch"*, then a dark interval *"J"*, which latter is followed by a brownish-yellow area *"C"*, containing large vacuoles. These vacuoles are in a single lens zone, which is the first lamellar surface of the lens of rabbits. Note the diagonally horizontal manner in which the vacuoles are distributed, due to the *anterior lens fibres*, which extend in the same direction. (An analogous grouping of vacuoles in relation to the fibre direction in human lenses is shown in Fig. 198.)

Fig. 284. Anterior cortical cataract produced by ultra-red and red rays in a half grown brown rabbit.

Eight weeks after exposure.

Total iris sphincter paralysis as an after effect of the experiment. Low magnification.

The anterior cortical cataract is composed of white scales and stripes in the area of the first lamella, while the surrounding cortex anterior and posterior to it is more or less clear.

It must be mentioned that the *anterior suture*, which in rabbits is nearly vertical in direction (see text to Fig. 178) is an area of predisposition to changes and is visible as an opaque stripe.

The *posterior*, somewhat yellow cataract is seen to the left of the anterior white opacity in the illustration. The iris is retracted above, so that the zonule becomes visible. The latter has been absorbed in its greater extent. A few remaining fibres are covered by pigment. The lens owing to the absorption of the zonule, is drawn away from the ciliary body, so that a flattening of its border can be observed. Iridodialysis is seen at this place.

In the area of the posterior cataract, there is a large deposit of pigment debris on the posterior capsule.

On the upper cornea there is a pterygium-like deposit of tissue. Earlier stages of this case of cataract are shown in the chapter pertaining to the iris Figs. 325 to 328.

See the same page, regarding the pigment changes of the iris.

Fig. 285a, b, c, d. Complete cataract by exposure to ultra-red and red rays. Duration of exposure three hours. Right lens of a black rabbit.

Examination two days after experiment.. The cornea shows a thin veil-like opacification (Fig. a). A few days later a parenchymatous keratitis developed, which has completely disappeared. Fig. b shows the other normal eye of this animal.

Fig. 285c illustrates an early transcient stage of the cataract. "d" shows the vacuole formation near the anterior capsule. A few weeks later the lens was completely opaque.

Figs. 286 to 291. Hypermature cataracts,

Fig. 286. Folds in the capsule and cholesterin crystals in a shrinking cataracta complicata.

Oc. 2, Obj. a2. Mrs. J. age 78, right eye'blind for many years. Intensively white cataract. Projection questionable. Slightly atrophic iris, no iridodonesis. The pigment border of the iris is absent.

The mottled white marble-like, very likely in part calcareous cataract, shows iridescent coloured cholesterin crystals in the peripheral cortical areas, which are arranged in a regular manner.

The radial concentric distribution of the groups of crystals in the anterior cortical periphery suggests the location of the suture areas. (Compare also the arrangement of the fissures filled with fluid, Fig. 209, also the brown-red foci in siderosis Fig. 278. This manner of arrangement according to our conception is due to the predisposition of the sutures to cataract involvement. It is also examplified by a similar type of involvement seen in the rosette-form changes in cataracta complicata.)

The anterior capsule is folded in several characteristic ways. I found these folds in this and in other cases to be 0,05 to 0,1 mm wide, extending in a worm-shaped manner and often showing a branching. As the lens is intensely white at the time these folds appear, the reflecting lines of the latter do not contrast greatly in focal light. *The capsular folds are therefore best observed by means of indirect illumination.*

These clinically unproven capsular folds I found present quite often in hypermature cataracts, especially in cataracta complicata, when observing with the slit-lamp. Their presence proves a shrinking of the lens substance. Not only have I found these folds in old calcareous cataracts, but also at times in the socalled soft juvenile cataracts of complicated types. They show blue by direct light. In cases where there is a dissemination of iris pigment, the latter shows a predisposition to attach itself to these capsular folds, so that brown stripes are formed.

Fig. 287. Gravitation of the nucleus in hypermature cataract (softening of the cortex). Mrs. B. B. age 68.

Case of Fig. 208 but the left eye. In focal light the lower cortical area shows the characteristic relief image of the axial nuclear surface, near the capsule. A Y formed suture (seam) is seen. The nucleus, owing to its weight, has sunken down. By turning the head to one side, the nucleus will gravitate in the corresponding direction. The displacement of the middle of the nucleus from the middle of the pupillary area is 2 to $2^1/_2$ mm.

The opening of the anterior capsule discharged a milky substance, and the nucleus was seen lying in the lower part of the capsular bag. After extraction it measured nearly 5 mm in thickness, and it had an equatorial diameter of 7 mm.

In the normal lens the nucleus is also of an increased specific gravity, compared to that of the cortical substance. We have proved this by the fact that we were able to centrifuge nuclear particles in cortical substance[46]).

Figs. 288 to 291. Secondary cataract.

By examining with the slitlamp secondary cataract showed a series of clinical peculiarities and characteristics with which we were insufficiently familiar or which were unknown.

Not alone are very thin cataractous films shown, which are invisible by all other methods, but the *vitreous* may also be clearly seen through openings in this membrane. After *discission* the vitreous *often* prolapses, hernia-like, into the anterior chamber. Fibrelike proliferations of the vitreous may at times be seen extending forward and connecting with the wound of entry of the needle in the cornea.

The prolapsed vitreous is often slightly clouded and its supporting structure, owing to contraction, and on account of deposits which lie on it, appears inc eased in luminosity. It practically always shows deposits of brownish and red dots.

The capsular membrane as it is left after extraction is often attached in places to the pupillary border. It is of irregular thickness and varies greatly in *tension* in its several areas.

This irregular tension causes *folds due to traction* (see Fig. 289) which show double parallel reflex lines. By the instillation of atropine and pilocarpine these folds may be modified. The membrane, in the light of the slitlamp, shows vivid interference (dispersion) colours (Figs. 288 and 289), green and red especially predominate.

This iridescence is only visible in the respective reflecting zone, which makes it easy of observation on account of the many folds in the membrane. The reflex lines thus produced may be accepted as linear reflecting areas (Vogt[28]).

Quite frequently secondary cataract shows clear spherical structures, the nature of which is still in question (Fig. 288).

Often these spheres are arranged in a string, like frog-eggs. Are these myelin droplets a proliferation of the epithelium? They may be seen singly inserted in the membrane, or appear to interrupt it.

Elschnig[140]), who first saw these spherical structures, describes them as being regenerative changes of the epithelium.

Fig. 288. Secondary cataract. K. Z. age 16. Lens removed 13 weeks ago (lamellar cataract, lineal extraction following discission).

Oc. 2, Obj. a2. To the right below a posterior synechia of the iris attached to a frog-egg-like conglomeration of clear spheres. The latter in the upper part of the mass are somewhat smaller and opaque. To the left a white opaque area (degenerated lens fibers?). Extending from this is a delicately folded iridescent capsular membrane. Within the latter there are two large clear spheres, which seem inserted into the membrane.

Fig. 289 a and b. Secondary cataract with folded iridescent cataractous membrane.

Case of Fig. 197. Six weeks after lineal extraction of traumatic cataract (Oc. 2, Obj. a3).

In Fig. a below, remnants of degenerated lens fibre, within these, white flakes. (Fig. b shows these characteristic cataract remnants.)

Above this the secondary membrane (posterior capsule?), with its radial folds due to traction. Note the iridescent coloured double reflex line. Farther up one may see pigment and cataract remnants, adjacent to the vividly iridescent secondary membrane.

Fig. 290. Dense folds of the capsule in secondary cataract. E. F. age 13.

(Injury by arrow 6 months ago.) Oc. 2, Obj. a2. Cataract following bulbar contusion. Partial iris adhesions. Spontaneous resorption. Note the pupillary border adherent to the cataract remnants, it shows a localized round dilatation at one place, due to the action of atropine. Along the pupillary border there is a bundle of dull white folds, irregularily curved downward, and attached to the iris. Pigment debris and a few blood-vessels pass diagonally over these folds and prove that a delicate exudative membrane covers these cataract remnants.

Fig. 291. Capsular folds in secondary cataract remnants. L. Sch. age 12.

Secondary cataract following traumatic cataract. (Perforation by wire $1^1/_2$ years ago, spontaneous resorption.)

In the central pupillary area there are dense secondary cataract remnants of the type shown in Fig. 289b. Over these white remnants can be seen numerous folds of the capsule, which show branching.

In this case as well as in other similar cases the folds as well as the double reflex lines do not contrast well on the luminous background with direct focal light. The folds are seen more distinctly with indirect illumination.

F.

EXAMINATION OF THE IRIS

In describing the examination of the iris with the slitlamp it is necessary to again draw attention to the sharp distinction between observing in *direct focal light* and by *transillumination*.

For observing by transillumination there are three methods at our disposal.

1. The iris tissue adjacent to the area which is to be examined is illuminated with the light directed at a flat angle. The light reflected from the luminous deep parts through the area to be examined is thus utilized.

2. The light may be projected through the pupil into the lens. The latter especially in age shows an increased internal reflection and fluorescence and this light is used to transilluminate areas situated laterally to it. In cases of aphakia this method of indirect illumination is less useful (compare *Staehli*[12])[13]).

3. The iris may be illuminated and transilluminated through the sclera. It is dependent on local conditions as to which method is of the greatest advantage. The indirect method is especially applicable in cases of atrophy and perforations and suggillation of the stroma (ecchymosis)*. At times cysts and tubercles may be discovered which are invisible by any other clinical method. The *sphincter pupillae* will show up as a sharply bordered plastic mass, by transillumination, under favorable optical conditions, especially in old age in cases of atrophy of the pigment layer and stroma, in light colored iridis. (Compare *Krueckmann*[128]).) Illustrations showing this have not been made for this issue of the atlas.

We are showing here the interesting relation of the size of the width of the sphincter to that of the pupil. In an individual aged 62, the sphincter mass was 1 mm in width when the pupil was 3 mm in diameter, $^4/_5$ when 5 mm, and $^3/_5$ when 8 mm in diameter. Therefore the indirect method of illumination is of great practical value in examining the iris.

Direct illumination is of value in examining the relief image of the iris and pupillary border, for the purpose of discovering scattered pigment, for illuminating the angle of the anterior chamber and observing the pupillary reaction. It is especially suited for determining *microscopic* partial or complete pupillary reaction. (For instance in tabes and paralysis.)

Especially valuable is the microscopic determination of beginning posterior synechiae, for instance in early increase of intraocular tension in cases of chronic or subacute iridocyclitis. There we may use miotics as long as the pupillary border remains free of attachments. I examine once or twice daily with the slitlamp under a 24 time magnification to determine if the pupil is round and free. Regarding the diagnosis of very incipient iridocyclitis (the incipiency of dew-like deposits and individual cell precipitates) see the chapter pertaining to the cornea.

---

* A case of *suggillation of the stroma by blood*, as I have observed it in contusio bulbi, Vogt[11]) could not be shown in this atlas. In focal light ecchymosis is not visible. An area of this kind presents a *vivid blood-red* color if observed by transillumination, when the bundle of light is projected to its side. To discover an ecchymosis in this manner may be of diagnostic value.

Fig. **288—297.**

Tafel **31.**

286

290

289 a

289 b

291

292

293

295

297

294

296

Verlag von Julius Springer, Berlin.

1. THE NORMAL IRIS

Fig. 292. Normal juvenile pupillary pigment iris border. Mr. St. age 22.
Left eye.

The right eye of this patient is shown in Fig. 304. Oc. 2, Obj. a3.

With the pupil at a medium dilatation the pupillary pigment border of the iris as seen from the front averages 0,06 mm in width. The pigment border is thrown into ruffled folds. These vary in width and thickness at different places. In certain areas they may be projected forward or flattened. The pupillary pigment border may be entirely absent for a certain distance as may be seen in Fig. 294, *without the condition being necessarily pathologic.*

This fact is especially important in the diagnosis following injury, as an inexperienced observer may in such cases presume that he is observing a ruptured sphincter. In judging the width of the pigment border it is important to exactly determine the direction of the illumination. As the surface of the pigment border descends in a steep manner toward the lens surface, it is more distinctly seen if it is illuminated vertically than tangential.

In the latter case it would not be possible to see that part of the border which is on the side toward the lens capsule. Therefore the temporal border should be illuminated from the nasal side and vice-versa (see *Hoehmann*[96]). Under *mydriasis* the width of the pigment border is greatly decreased, in fact it may be totally covered by the small circle of the iris, so that it may appear to be absent. Therefore the pigment border should not be examined under mydriasis.

The average width of the border in its anterior direction in a series of young individuals, with a pupillary diameter of 3,5 mm, varied from 0,04 to 1,1 mm. Exceptions to this average are common.

In this case a light colored iris is illustrated. To the right the pigment border is shown in focal, to the left by transillumination in the light diffusely reflected by the lens. In the upper part of the picture, the white light reflex of the lens may be seen.

Fig. 293. The normal iris border pigment in old age. Mr. H. age 65.

Bilateral miosis. Right eye. Oc. 2, Obj. a2.

In old age I have often found a general *widening* and *thickening* of the iris border in some areas. Its width can be several times as great as is seen in youth. *In addition there is often an angular rough edged irregularity, so that we may speak of knobs rather than of swellings.* (Compare for instance Fig. 298.)

In the case of Fig. 293 the swellings are somewhat pronounced, as the pupil is only 1,5 mm in diameter.

Immense swollen mounds are seen in this case, especially above and upward in the temporal direction. Several project quite forward, compared to their average elevation. In addition there is a senile scattering of pigment, which according to our observations is always seen in old age (see *Hoehmann*[96]). The high degree of depigmentation and wandering of pigment which was construed by *Koeppe*[97]), to be a symptom of threatening glaucoma (pre-glaucomatous), seen in old age, could not be so accepted by us, in view of the results of the experiments we made to determine if this were so. Also note the irregular prominence of the stroma continuations toward the pupil, which cause a slightly irregular angular pupillary form.

8*

Although we may at times see a broad pigment border in youth, a comparative study in a large number of individuals has convinced us, that we are justified in speaking of a *senile widening* of the *pigment border.* The latter, in all probability is always accompanied by a scattering of pigment.

If we inquire as to the cause of this "hypertrophy", we may at first be inclined to think of a proliferation of the pigment in the pupillary border. There are however no reasons for accepting this view. The following facts are to be considered:

The pigment border shows its greatest width in miosis, naturally for the reason that in miosis the pupillary circumference is decreased in proportion to its diameter. In a pupil 2 mm wide the circumference measures about 6 mm, and in one 4 mm wide, about 12 mm. One would expect therefrom that the reduction of the pupillary diameter in age would cause a widening of the pigment border. Measurements however have shown that the latter is much in excess of what may be expected from the miosis alone.

What is also of importance in *artificial* miosis is that the increase in width of the pupillary border is in excess of what may be explained as being due to a decrease in the pupillary diameter.

For example: Mrs. J. age $74^1/_2$. Pigment border of one eye when the pupil is 4 mm in diameter is 40—60 microns. Its continuity is interrupted in places. By artificial miosis (pilocarpine-pupillary diameter 1,9 mm) the pigment border measures 0,125—0,16 mm in width *in certain areas much wider*, up to 0,5 mm. Extra wide areas of the type suggest an *eversion* due to traction.

In children and in youth this excessive widening due to miosis is less frequently seen. It follows from these experiments that the widening of the pupillary border in the miosis due to age is the result of an *eversion* of the pigment surface caused by a senile rigidity of the stroma, especially the vessels. In the contraction of the sphincter this stroma which is anterior to the sphincter is not able to follow sufficiently and the pigment border therefore is everted. I could often determine that the miotic pupillary border failed to rest on the lens capsule. An interval may be present. I also have frequently found a similar separation of the pupillary area of the iris from the lens capsule in non-miotic pupils. If one often studies the radial folds of the retinal pigment layer with the slitlamp, one will find that the law of the "physiologic pupillary closure" is not necessarily a certainty. It is possible that normally there are communicating areas between the anterior and posterior chambers.

Fig. 294. Normal pigment border repeatedly interrupted in one area. Mrs. W. age 52.

Oc. 4, Obj. a2. The pigment border shows no pathologic changes. In a large area it appears segmented, that is separated into several parts of similar length. At one place there are only small particles of pigment debris, at another the pigment border is entirely absent. I have often found this condition in normal eyes especially in older persons. These defects are congenital in some cases, in others acquired. Regarding the latter deposits compare Figs. 300 to 314.

Fig. 295. Retinal pigment layer of the iris near the pupillary border seen through a translucent normal light colored iris. Patient aged 16.

Oc. 2, Obj. a2. The tissues anterior to the retinal pigment layer near the pupillary border (axially of the circulus iridis minor), are thin and colorless. In focal light therefore the pigment shows yellowish-brown to green, through it. While in this case the iris appears blue because of an increased reflection of short-waved rays, the border near the pigment collar has a yellowish tinge given it by the pigment which shows through it. The same condition is also seen in Fig. 292. I have often found this relation in light colored irides.

Fig. 296. *Normal iris with the retinal pigment layer visible in circumscribed areas.*

In an area near the pupillary border the connective tissue stroma of the bluish green iris of Mr. L., aged 24 years, is quite thin, in places it is apparently absent, and the retinal pigment is exposed to view. Note the brown triangular spot in the center of the illustration. The white fibers of the iris stroma are practically absent. Oc. 4, Obj. a2. This condition is probably congenital and not at all rare. (Compare *Koeppe.*)

Fig. 297. *Normal iris with the retinal pigment layer visible in a crypt of the iris.*

Oc. 4, Obj. a2. At the bottom of deep crypts especially in light colored irides one may often see the retinal dark-brown pigment layer. The crypts shown in the illustration are in the iris of a woman 50 years of age with a normal light bluish-gray iris.

Fig. 298. *Senile pupillary border with a knob-like angular pigment border and a homogeneous structure of the adjacent stroma. Mr. B. age 80.*

Oc. 4, Obj. a2. Right eye. Light colored iris, scattered pigment above. Note the angular form of the border and the homogeneous (hyaline?) character of the stroma striping in the vicinity, through which latter the pigment layer is slightly visible. The relief of the iris seems flattened.

Fig. 299. *Fibers and strands of the normal iris stroma.*

The pupillary membrane in fetal life is not alone supplied by blood from the vessels through the vitreous, but also from the iris root, which exists at that time. The manifold strands and fibers which we find forming the iris stroma in almost all eyes are the remnants of these vessels. We may find fine spider-web like fibers, at times attached to the lens capsule. More often tough strands in the form of the trabeculae, with tips and cone-shaped projections extending forward into the aqueous are seen.

These latter may be taken for tubercles if not carefully examined. In this case of Mr. M. aged 32 years, there are bridge-like fibers which give design to the stroma, that is that part of the iris which in fetal life formed the pupillary border.

The pupil is dilated, so that the pupillary border is partly covered by stroma and the bridge-like trabeculae are drawn anterior to the pupillary edge.

2. THE PATHOLOGIC IRIS

It is difficult to distinguish between senile and pathologic changes in the iris.

Figs. 293 and 298 show tissue changes which may be considered pathologic, were they not so regularly found present in age. Atrophy of the pigment border and scattering of pigment, which occurs in all individuals in age is shown in the next following illustrations. Regarding the atrophy and scattering of pigment compare the reports of *Augstein* (1904 and 1912[54]), *Goldberg* (1907[145]), *Krueckmann* (a coloured illustration in 1907[123]), *Vossius* (1910[94a]), *Axenfeld* (1911[95]), *Hochmann* (1912[96]), *Staehli* (1912[12]), *Koeppe* (1916[97]) Migration into the stroma), and others. The senile scattering was first seen by *Goldberg* and *Hoehmann*. Personally I can assert that senile depigmentation occurs sooner or later in all individuals.

In cases where this depigmentation is of a high degree it need not be considered a symptom of threatening glaucoma, as has been asserted. In determining atrophy of the pigment it is important to differentiate between observations by focal and in indirect light. The latter is especially valuable in these. investigations.

Figs. 300 and 301. Atrophy of the pigment layer and connective tissue stroma.

A) OBSERVATIONS IN FOCAL LIGHT

Fig. 300. Atrophy of the retinal pigment layer near the pupil. (Rarefaction of the pupillary pigment border.) Mrs. N. 65 years.

Oc. 4, Obj. a2. In direct focal light we may observe that the pupillary border is absent over an extensive area. In one place there is only a thin transparent glassy membrane. One is impressed by the fact that this thin membrane must be the depigmented foundation which represents the "supporting framework" of the normal pupillary border, *still in its original form*, after all or the greater part of the pigment has disappeared or been carried away. In some of these cases (Fig. 301) one may see particles of pigment. debris attached to or suspended from the remaining framework of the pupillary ·border.

Note in addition the *brown tissue areas* adjacent to the depigmented zones. They are often found present in these cases (compare Fig. 301). These yellowish brown tissue areas are *only found* in the vicinity of depigmented zones. so that it appears quite probable that they have received their coloration from the pigment which has undergone transit. This is well illustrated in Fig. 300. Above, in the vicinity of the still pigmented pupillary border zone there is no area of yellowish brown coloration created by pigment in transit, while below a very pronounced area of this kind may be seen. Pigment carried into the tissues of the iris is best seen by transillumination. In focal light it is often invisible.

Regarding this senile depigmentation we may establish two directions, by means of the slitlamp, in which the pigment is carried away.

1. Through the aqueous, the pigment being deposited in small particles and dust-form on the iris, lens capsule and corneal surface.

2. By being carried along the stroma of the iris in the manner described above. (*Koeppe*[97]).

Fig. **298—307.**   Tafel **32.**

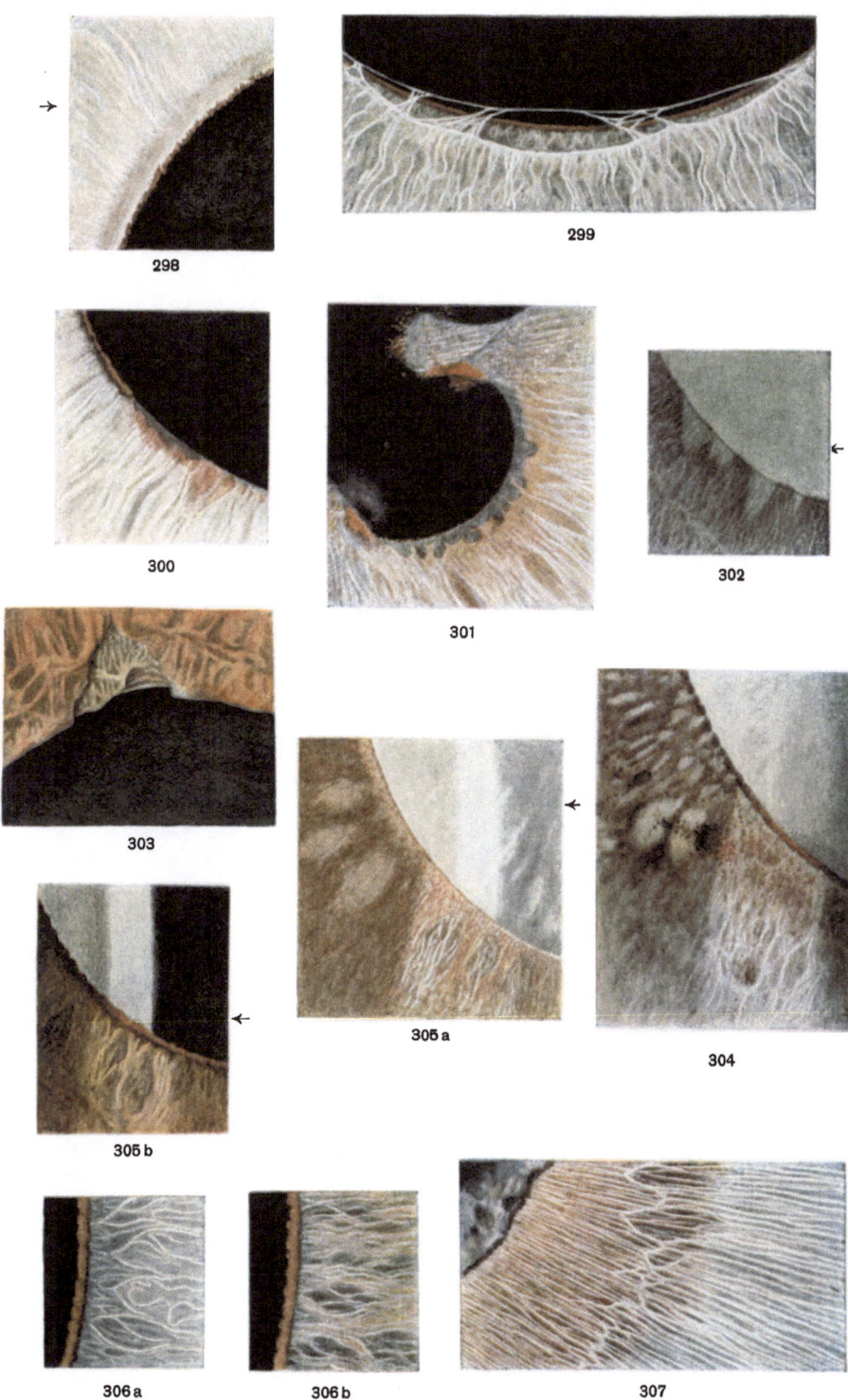

298

299

300

301

302

303

305 a

304

305 b

306 a

306 b

307

These observations lead to the conclusion that this *primarily* senile process, as it may step by step be observed with the slitlamp, is due to a lowering in vitality, respectively a gradual death of the pigment epithelial cells in the pupillary border of the iris.

Fig. 301. Atrophy of the retinal layer in senility and following iridocyclitis.

Oc. 4, Obj. a2. Sister F. K. age 69, has suffered intermittent attacks of iridocyclitis for the past thirty years in her highly myopic left eye. There are several posterior synechia, evidently of old standing. The pigment of the iris stroma especially in the vicinity of the adhesions is disappearing by migration.

As in Fig. 300 the pupillary border in certain areas is represented only by a thin membrane. It is composed practically of a hull with here and there a few adhering pigment particles. The yellowish brown pigment areas as also described in Fig. 300 are here seen in the vicinity of the transparent membranous framework. In this case there is a mydriasis due to the effects of atropine. It is also worthy of note that the yellowish brown areas seem to bulge forward, due to the dilatation, no doubt the results of a traction exerted by the dilating fibers.

B) OBSERVATIONS BY INDIRECT ILLUMINATION

Fig. 302. Atrophy of the pigment border and the adjacent retinal iris pigment layer. Mrs. S. age 84. Senile Cataract.

Oc. 2, Obj. a2. Indirect illumination (transillumination) *not alone* exposes an atrophy of the *pupillary border* but also of the retinal layer situated behind the pupillary border zone of the iris stroma. In this case the pigment with the exception of a few remnants has in certain areas entirely disappeared, so that the *pupillary iris border by transillumination* in large areas has an appearance *resembling tinder*.

The translucent areas have a sector formed shape with their bases toward the pupillary border. At the latter the pigment is entirely absent, while in the intermediate areas it is still partly present. The design of the surface of the iris is indistinct and the stroma presents senile atrophy.

Regarding the technic of observing this by *transillumination*, it must again be mentioned that the bundle of light should be directed onto the *adjacent lens surface, not onto the pupillary border itself*, and that the observation of the temporal border of the pupil calls for an illumination from the nasal side and vice versa. If this be kept in mind the observations may be quite easily made. As the reflection from within the lens is increased in *cataract formation*, the atrophy of the retinal layer will usually show in a more pronounced manner than if the lens were clear.

The iris tissue may possibly be quite translucent in large areas, so that it can be transilluminated from the anterior side.

We may then see the illuminated and transilluminated areas side by side, which will present a peculiar iridescence.

This in all probability explains the recent assertion of an author[98]), that he had observed mother-of-pearl like spots in the iris following attacks of glaucoma, which latter spots he designated as symptoms of a previous glaucoma.

It must be especially emphasized that in *direct* focal light one may only observe a rarefaction of the pigment of the pupillary border. The diagnosis of the disappearance of the retinal pigment layer, which is so typic of old age can only be shown by *transillumination* in the manner described. Further observations of atrophy of the pigment layer are illustrated in Figs. 222, 305a and 304.

Fig. 303. Rupture of the iris sphincter following contusio bulbi. Mrs. K. age 64 (case of Fig. 336).

The rupture extends into the superficial layer of the iris stroma. The pupil is slightly dilated due to the rupture and because of glaucoma. The pigment border *cannot* be seen. The edge of the tear is translucent and void of pigment. Ruptures of the pupillary border, some of old standing are shown in Figs. 277a and 336.

Fig. 304. The iris in cataract due to heterochromia.

In heterochromia as is well known the least pigmented eye is predisposed to cataract formation. The difference in color, according to our observations, need not be a pronounced one, as may be seen by Figs. 304 to 306.

Fig. 304 presents the iris in the case of Mr. St. age 22 (Fig. 292). Complete cataract, right. Oc. 2, Obj. a3. There are a few precipitates on the posterior corneal surface. The eye had always been free of irritation. The left eye is bluish macroscopically while the right cataractous eye is grayish green.

In Fig. 304, as in Fig. 292, the iris is shown by focal light to the right and by transillumination to the left. Both manners of illumination show scattered pigment outside and within the stroma.

Real rarefaction of the pigment layer and apparently of the stroma are seen only in indirect light. The pigment border is uninterrupted and in the stroma one may observe luminous defects, in part crypts. That the delicacy of the stroma has suffered can be seen by comparing the area illuminated by direct light in this figure with that of Fig. 292.

Fig. 305 a and b. Cataract in heterochromia. Mrs. F. K. age 56, Fig. a right iris.

For years the vision in this eye has been reduced in acuity. There is a complete cataract. The right part of the illustration shows the iris in focal light, the left by transillumination. Oc. 2, Obj. a2. Fig. b presents the left normal iris of this case, also in focal light and by transillumination. Oc. 2, Obj. a3. (The lens of this eye presents a peculiar spotted cataract, probably congenital in origin, which is illustrated in Fig. 232.)

The healthy iris (Fig. b) is brown in color, the richly pigmented border is intact. The pathologic iris (Fig. a) has a yellowish gray to greenish red color, and the pigment border is absent. The whole of the retinal pigment layer of the iris seems rarefied and the stroma is almost free of pigment and it is translucent in some areas.

Fig. 306 a and b. Normal and pathologic sphincter areas in a case of heterochromia. Cooper L. age 41.

He had noted a gradual reduction of visual acuity of the right eye for some years. In his military service record of twenty years ago, the visual acuity of this right eye was recorded as being 1, and he reported that this eye served him well when at target practice. As far back as he can remember the right iris was lighter in color, as compared to his left.

The posterior corneal surface presents many colorless clear mostly star-formed precipitates. The right iris is light blue, (Fig. a) the left some lighter in color (Fig. b), with individual superficial chrome-yellow areas of pigmentation. Under a magnification of 37 times (Fig. a), the tissues of the pathologic iris are somewhat rarefied. This rarefaction manifests itself in the absence of the wooly dense character of the stroma as seen in the normal left iris. The white lines of the sphincter vessels (Fig. a) are extraordinarily distinct and luminous, while in the normal stroma (Fig. b) they are less distinctly visible, being covered by wooly feltlike tissue. Their walls for this reason do not especially reflect the light. The pigment border in both forms of illumination shows no changes, which may definitely be accepted as such. The lens shows a uniform, axially somewhat more dense opacification of the superficial posterior cortex. It is of a delicate porous character. There is an extensive iridescence of the posterior reflecting zone and in certain areas large cholesterin crystals may be seen. Anteriorly there are a number of axial sub-capsular vacuoles.

The vitreous, wherever it is visible, shows a clouding of the supporting framework and many coarse white punctate deposits. The left eye presents no peculiarities.

Fig. 307. Atrophy of the iris tissues after many years of iritis.

Oc. 2, Obj. a2. Focal illumination. F. age 55. He has for years had a bilateral chorioretinitis and iritis due to lues. Owing to seclusio pupillae of the right eye an iridectomy was done a year ago. In the illustration the iris tissue of the pupillary border is attached by a dense exudative mass to the lens capsule. The stroma is reduced in quantity, atrophic and as a result of traction due to the shrinking the iris trabeculae are quite *tensely drawn* and therefore parallel to one another. This area on account of the traction is radially extended. Here one may note the translucence of the retinal pigment layer through the rarefied stroma, especially in the area of the crypts.

Fig. 308 and 309. Ectropium uveae acquisitum in glaucoma absolutum, complicating sarcoma choroideae. Case of Figs. 14 and 16.

The iris stroma is snow white, evidently a connective tissue change in which the radial trabeculae strands are still discernable. (Fig. 309 presents a lower area of the iris magnified 24 times.)

A part of the visible blood vessels, which show a radial and concentric formation are in all probability newly formed. It is certain that the veins are greatly dilated. I have observed similar evidences of stasis in combination with atrophy of the iris stroma in other cases of absolute glaucoma.

A few posterior synechia are seen below. It is peculiar *that just at these areas* the ectropium uveae is comparatively narrow, which suggests that the latter is a result of *scar contraction of the stroma*, so that a kind of a flap is formed.

In addition to this scar contraction there is also an active proliferation of the pigment epithelium on the anterior surface of the iris, this is according to *Siegrist* and *Stern*.

Fig. 310. Meridional section through the ectropium uveae of Figs. 308 and 309.

The retinal pigment layer extends onto and over the anterior surface of the iris. The latter forms an extensive anterior peripheral synechia with the cornea. All of the sphincter is everted forward, carrying with it individual *Koganei's* clumped cell masses. The anteriorly everted retinal pigment layer is *rarefied* by *stretching* and its physiologic folds have been thereby eliminated. Between the pigment border and the cornea there are extensive flat veinous sinuses directly under the surface of the iris stroma as is shown in Figs. 308 and 309. (The veins on the surface of the melanosarcoma of the choroid, which may be seen directly posterior to the lens pole are also ectatic.)

Fig. 311. Atrophy and changes in the coloration of the iris following herpes zoster ophthalmicus. Mrs. M. age 60. (Case of Fig. 45.)

Four months after herpes zoster ophthalmicus. Right eye. The iris in parts is adherent. Macroscopically it is of a pale grayish blue. Temporalward it is brown. (The other normal iris is of a different light blue color.)

In a temporal area of the iris there may be seen deep round pit-like excavations, in which large masses of pigment are deposited. The areas appear irregular, depressed and their surfaces are smooth as if covered by a layer of collodion. In some places there are newly formed blood vessels. Note in addition the row of knot-like prominences in the outer part of the illustration. Their *apices* are free of pigment. Regarding the characteristic corneal maculae of this case compare Fig. 45.

At the time when the illustration was made there were individual precipitates up to 0,15 mm in size, with pigment and star-shaped borders, as well as folds in Descemet's membrane.

Fig. 312a and b. Pigment globules. Mrs. B. age 68.

Following an iridectomy preliminary to the extraction of a senile cataract there was present a bulbar irritation lasting for several weeks, without evidences of keratitis or iritis. (After the extraction it failed to recurr.)

When the irritation subsided the two pigment globules, shown in Fig. a, under a 24 time magnification, were observed. They had a slightly flattened base, one was on the pupillary border of the iris, the other was on the lens capsule. Fig. b presents one of these globules under a higher (37 time) magnification. They are smooth and practically round. The lower one after the extraction of the cataract was found intact at its original site.

Fig. 313 Scattering of pigment in senility and in absolute glaucoma. Mrs. Sch. age 71.

Oc. 2, Obj. a2. Left eye, absolute glaucoma. The scattering of pigment, first described by *Hoehmann*[95]) in 1912, and later considered suspicious of a glaucomatous tendency and pre-glaucomatous, by *Koeppe*[97]), according to my observations is practically always found present in old age. It is more pronounced, at times after contusio bulbi, of a high degree in addition to the migration of pigment in glaucoma*, in iritis, and especially following operative interference with the anterior chamber, if the iris was damaged (*Augstein, Vossius,* and others). The pigment which is disturbed and washed away, attaches itself to the projections of the iris, as may be seen in Fig. 313, where the pigment dust is freely deposited on a balcony-like projection of the iris. There are also present remnants of posterior synechia and the pigment of the pupillary border is diffusely scattered. By superficial examination the socalled *clumped cells* of *Koganei* may possibly be confused with scattered pigment. They are especially found near the pupillary border. (*Elsching, Lauber*[99]) also *Koeppe*).

Fig. 314. Scattering of pigment following contusio bulbi.

Traumatic subcapsular cataract. Oc. 2, Obj. a2. Mr. A. F. age 45. Undilated pupil. Six months ago he sustained a bulbar contusion of the right eye, by a piece

---

* The scattering of pigment in this case is probably of a secondary nature. That scattering of pigment, on the other hand, preceeds glaucoma (*Koeppe*), has not been confirmed by us in any case. Some eyes presenting an extreme amount of pigment scattering, showed normal tension, ad endured medicinal mydriasis.

Fig. 308—315.

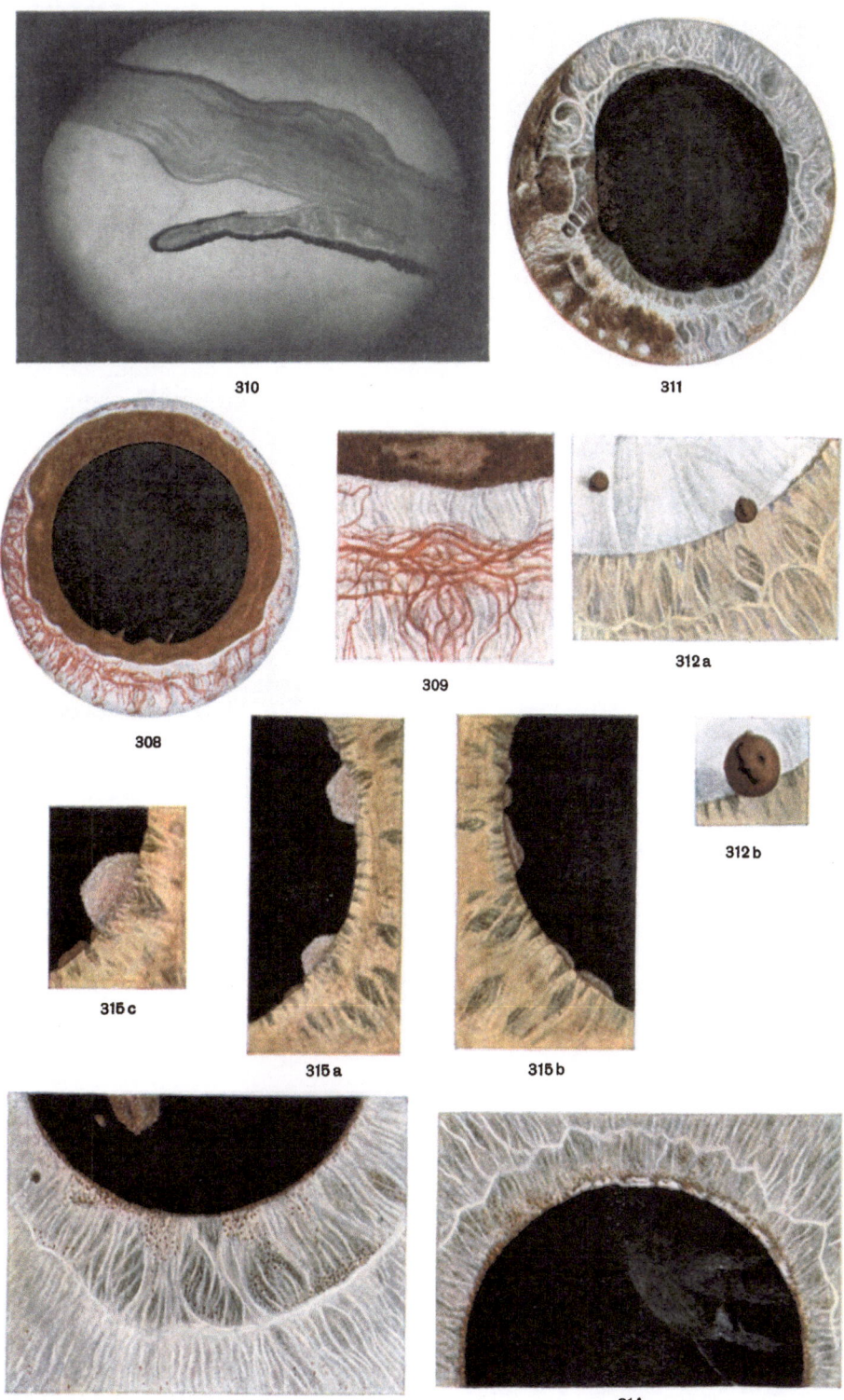

310

311

309

312a

308

312b

315c

315a

315b

313

314

Verlag von Julius Springer, Berlin.

of wood. Today $VisR = ^6/_8$ (Hm 2,5 D), $VisL = 1$ (Hm 0,75 D). In the right eye there is an anterior subcapsular traumatic cataract visible only by slitlamp illumination.

In the superficial and subcapsular cortex there are striped and grayish white spider-web-like opacities of a slight somewhat varying density, which extend over the whole inner upper pupillary area of the lens. By medicinal dilatation of. the pupil we may further observe that they extend peripheralward in the lens, increasing in density. and showing linear fiber striping. The cataract is situated in a flat extremely thin layer of the lens. By utilizing the posterior lens as a source of light for the purpose of examining the opacity by *transillumination* we may note that the opacity is composed of minute droplets. The shagreen over the opaque area shows a slight iridescence. Here and there may be seen a delicate fixed irregular spot on the lens capsule just anterior to the opacity, as well as a few granules of pigment. The *pupillary pigment border* in the inner upper third of the pupillary margin is defective or absent. The pigment is scattered irregularly over the adjacent stroma. Instead of the pupillary pigment border there are individual *white* tissue areas, however no discernable ruptures of the sphincter.

Below there are no changes in the pigmented pupillary border, nor does the lens present opacities. The right pupil is slightly dilated when compared to the size of the left. In the vitreous there are no changes which may definitely be diagnosed as deposits, however there are many pigment dots on the posterior lens capsule.

The left eye presents no peculiarities.

This case illustrates:

1. That a post-traumatic scattering of pigment and destruction of the pupillary pigment border, *without rupture of the sphincter*, may occur following contusio bulbi.

2. That contusio bulbi may at times be followed by a thin flat anterior subcapsular traumatic cataract, which may remain stationary (compare also the text to Fig. 277).

Figs. 315 and 316 a, b and c. Tubercles of the iris in chronic iridocyclitis.

In chronic iridocyclitis with precipitate formation there are seen quite often fungi-like suddenly appearing growths, compare *Krueckmann*[123]), *Gilbert, Koeppe* and others, reminding of bovista (Lycoperdon gummatum), which latter just show the cupula of their seed capsule above the carpet of the forest flora. Just as they suddenly appear, these growths may within days have again disappeared. As a rule these solid tubercles occur on the pupillary border, causing adhesions with the lens capsule, if they are in contact with it.

If one dilates the pupil at this time, one may note a deposit of exudate and pigment on the lens capsule, at the former site of the tubercle (Fig. 315).

At times these tubercles may unite to form a transcient (regarding transcient iris tubercles [tuberculides?] compare the observations of *Stock*[131]), *Krueckmann*[123]), *Igersheimer*[132]), *Koeppe*[134]) [slitlamp], *v. Hippel*[134]), and others, especially the work of *Gilbert*[130]) continuous border, which latter may cover large areas of the pupillary border, resembling a lawn.

In other cases these tubercles are found in the anterior stroma or they *appear from under the pupillary border* as in Fig. 316 b. Surrounding the tubercle, especially if the latter has existed for some time, one may observe dilated or newly formed vessels (Fig. 315). As a rule however the surroundings are unchanged. The surface of the tubercles is not a smooth one, hence the comparison of it to the bovista, which also show delicate crumbly tops. There may also be melanin granules on the tubercle surface. The crumbly appearance of the surface is probably due to cell elements.

Fig. 316 shows a tubercle of the case of Fig. 20, 23. Miss S. age 25 who had a bilateral chronic iridocyclitis for the past three months. (Wassermann negative, Tuberculin test uncertain.)

Fig. 316a to c are tubercles of Miss M. age 60 who in the past six months suffered intermittent attacks of (bilateral) chronic iridocyclitis. There are dewlike changes in the endothelium without precipitates. A ciliary injection has not been observed. The supporting framework of the vitreous shows fine white punctate deposits. Vision = 1 in both eyes. The tubercle formation in this case was of a very transcient nature. Wassermann and tuberculin tests were not made.

Fig. 317. The iris in sympathetic ophthalmia (sympathizing eye). A six year old girl E. H.

$2^3/_4$ months after the onset of sympathetic inflammation. (The left eye sustained a perforation at the limbus by a knife cut four weeks previous to the onset of sympathetic inflammation in the right eye.) Low magnification. Case of Fig. 39.

On account of seclusio pupillae an "iris bombe" has developed. Dense pigment exudates on the anterior capsule and many tubercle-like mounds are seen on the iris, temporalward however there is a *hollow depression* of the relief of the iris. The iris contains many newly formed blood vessels. Precipitates are present on the cornea and there is a moderate ciliary injection.

Fig. 318. Attachment of the posterior surface of the iris to the lens capsule by a perforating scar. (Case of Fig. 20o.)

Oc. 2, Obj. a2. Perforation eleven years ago. The scar is glazed, zig-zag in form, and has extended through the tissues in a concentric direction. In its immediate vicinity there is a scattering of pigment. The radial trabeculae are only partly present over the scar area. The lens shows a gray superficial opacification. The capsule has a deposit of pigmented stars. The fixation at the scar is demonstrated by the reaction of the pupil. When the pupil is of medium size it is round, if contracted by light a sharp angular kinking occurs at the site of the scar (Fig. 318). At this time the iris stroma is rolled over until it covers the pigment border at the kinked angle. By artificial dilatation the synechia may also be demonstrated.

Fig. 319. Circumscribed congenital ectropion of the lower outer pupillary border. Mrs. G. age 61.

Oc. 2, Obj. a3. (Case of Fig. 215.) In an area 1,8 mm in length the iris pigment border is greatly increased in width — in places up to 0,3 mm, so that we may speak of an ectropion (congenital). Partial eversions of this kind are very common. Below there is in connection with this defect, an area 0,17 mm in length presenting scattering of the pigment. Fig. 215 shows the same case under mydriasis and with a lower magnification. It illustrates the manner in which an ectropion of this kind may be made to disappear.

Fig. **316—324.** Tafel **34.**

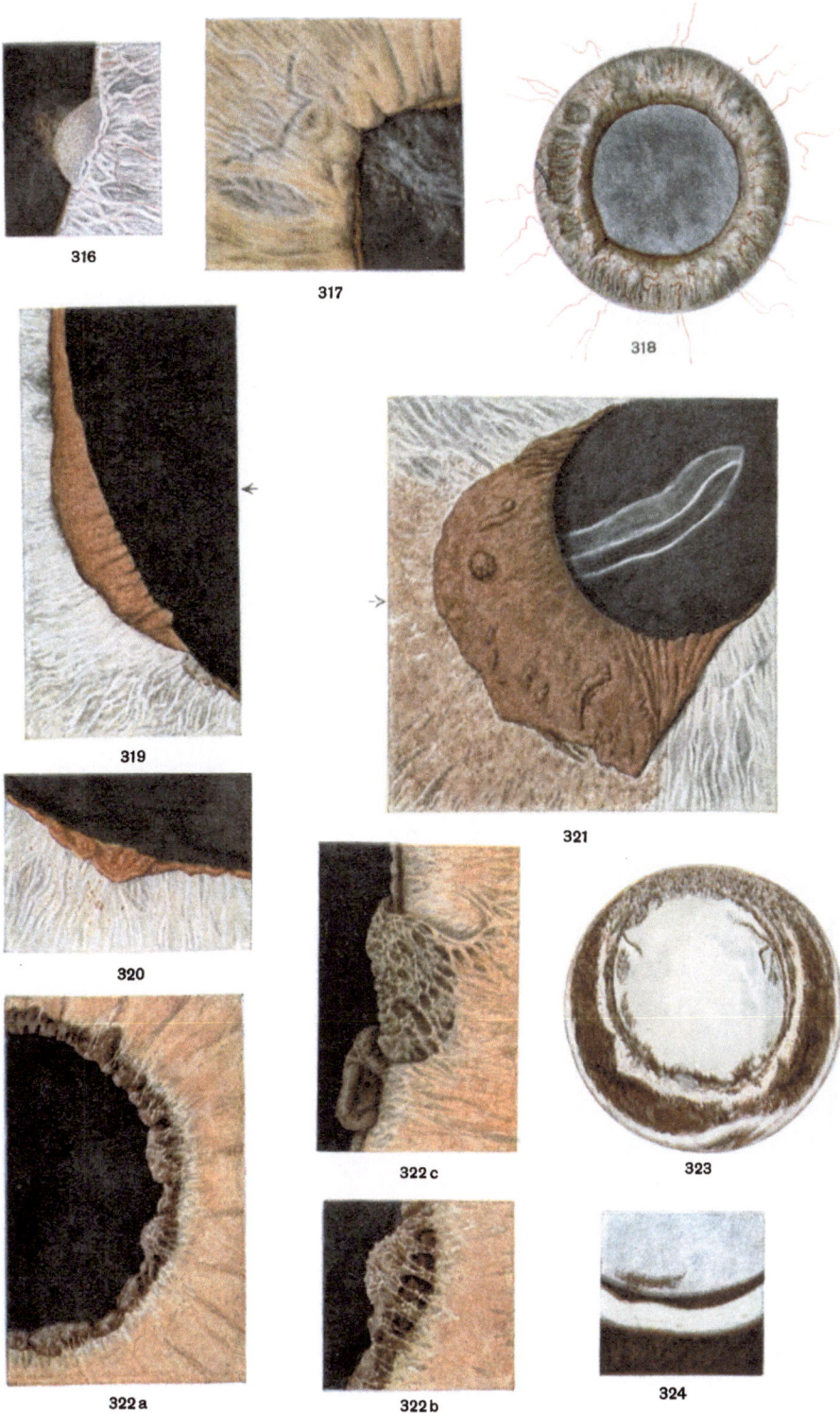

316

317

318

319

320

321

322a

322c

322b

323

324

Verlag von Julius Springer, Berlin.

Fig. 320. Circumscribed congenital ectropion of the retinal iris pigment layer, with scattering of the pigment. Mrs. F. age 74.

Oc. 2, Obj. a3. Lower outer pupillary border. The scattering of the pigment has occurred near the ectropion, that is it has involved the area of greater pigmentation. The ectropion measures 0,25 mm in its greater radial width. It appears drawn over the pigment border in an apron-like manner and I think it worthy of note that it contains radial folds, which are probably *folds of traction*. We may therefore conclude that we are dealing more with an *active drawing over* of the pigment, rather than an extension in growth.

Fig. 321. Apron-like congenital uveal ectropion. Mrs. T. age 67.

Oc. 2, Obj. a2. Right eye, temporal and downward. The ectropion in this case measures 2 mm in a radial direction. Notice the pigment folds of the border, which almost disappear on mydriasis. There are individual small pigment knobs and swellings. The pigment surface is in the same layer as the superficial iris stroma. The latter is less developed peripheral to the ectropion, so that the retinal pigment layer becomes visible. The balance of the pigment border is normal. In the anterior cortex one may observe the white edges of a fissure filled with fluid. Does an ectropion of this kind represent a defect in the stroma, or is the pigment layer everted by traction? It is possible that both may be factors.

In judging certain cases it is important to remember that the iris zone axialward of the circulus minor develops in the later months of fetal life. (This however is questioned by *Seefelder*.[101])

At this time the iris stroma and the pigment layer under it is displaced from the area of the circulus minor in an axial direction. This *extension* of the stroma and pigment layer evidently occurs in a regular symmetric manner, however the extension of the pigment layer may predominate, which creates the normal pigment collar of the pupillary border. If at this time, in a certain zone, for any reason there may be a retardation in the growth of the superficial stroma, a defect in the latter, exposing the pigment layer would occur *or* the pigment layer as such would be drawn over and everted, because the pigment border is held taunt by the pigment layer. A certain tension would arise, which would explain the development of folds in the pigment layer. Coinciding with a hypothesis of this kind is the fact that this form of congenital pigment ectropion is always situated axialward of the circulus minor, very seldomly involving the latter.

Fig. 322 a, b and c. Irregular knob-like bulging pigment border (Colsmann, Holmes, Hirschberg[126]) and others) in congenital attachment of the pigment border to the circulus minor. Mr. H. S. age 31.

This condition was incidentally observed at the time when a foreign body was removed from the left cornea. The pigment border in both eyes is of a peculiar form. It is composed of thick brown knobs and mounds of various sizes, which project into the aqueous to a height and width of 0,5 mm. They are individual as well as confluent. Fig. 322a shows the nasal pupillary border of the right eye under a 24 times magnification. The pupil is 4 mm wide. The balance of the pupillary border as well as that of the left eye is similar in appearance. Some of the knobs in the left eye are larger. Both eyes shown an active pupillary reaction. The lens capsule presents no evidences of remnants of the fetal pupillary membrane. The appearance of the pupillary border in Fig. a if superficially observed with insufficient illumination, would make it appear as if the pupil were irregular and angular in outline.

Fig. b presents an area of the right pupillary border under higher (68 times) magnification. The surface of the pigmented mounds is finely granular, as is considered normal, and the mound bulges peripheralward. The increased magnification shows fine yellow tissue fibers on the surface of the large mound, which represents the *extension of the circulus minor*. It continues net-like onto the inward eversion of the pupillary border. This is even better illustrated in Fig. c, which shows a temporally situated mound of the left eye (68 times linear magnification). The trabeculae here and at other areas not illustrated, are better developed and extend over the mound surface into a delicate dense network, which has the form and type of the balance of the stroma of the iris. The pigment mound under this covering is *wrinkled*, that is *everted and rolled inward*.

Fig. c shows that under the enormous pigment roll there is inserted from above the *normal* pigment border and stroma. The normal area however has only a length of about 1,5 mm.

The outward eversion of the pupillary border, *by its dense attachment to the circulus minor* makes it appear probable, that in fetal life a firm attachment exists between them, at a time when these two structures were still adjacent to one another. The extension of the axial stroma area which contains the sphincter had not occurred. On account of this firm attachment the retinal pigment layer could not extend in a surface direction, and as it was further developing, it was mechanically forced to advance between the circulus minor and the area of its attachment, hence the origin of its eversion and deformity.

Partial eversions of this kind are not rare, compare Fig. 142. This and the case just described leads one to conclude, that the traction of the stroma, not alone causes a deformity but also an eversion of the pigment layer.

I wish to further draw attention to the light-brown wart-like pigment clumps which are attached to the surface of the network of the stroma, which surrounds the pigment mounds. Similar pigment clumps are often attached to so-called pupillary membrane fibers (see Fig. 142), such as have been described by *Brückner*[60]). The structures just described may be classed with the so-called *"flocculi"*[114)][126)][127]).

Figs. 323 to 328. Experimental depigmentation of the irides in living rabbits by exposure to shortwave ultra-red, to which were added some outer red rays. (Vogt[93]*).)*

Compare the artificial development of cataracts Figs. 282 to 285.

In addition to the cataract, as a rule a few days later, a scaling off of the superficial pigment near the pupillary border was observed, so that a whitish ring, concentric to it appeared.

The detached pigment could be seen in clumps at the bottom of the anterior chamber. While the anterior pigment disappeared following this exposure, we noted clump-like *pigment proliferations* appearing under the pupillary border. A further peculiar development was a paralysis of the sphincter of the iris. The pupil, a few days after the exposure remained fully dilated and rigid. The cornea does not seem to be affected, at times it was slightly clouded, in one case extremely so. This latter case was followed by vascularization, later on the infiltration disappeared.

Pigmented hair surrounding the eye are often shed, the new hair growing at this site is *free of pigment.*

Figs. 323 and 324. These present the right eye of a black, several years old male rabbit, which was exposed for three hours. Regarding the complete cataract resulting from this exposure, compare Fig. 285. The depigmentation of the iris was quite extensive in an upward direction. The iris is thin and atrophic. On the anterior lens capsule near the pupillary border there are large pigment clumps-visible only at two or three places., There is a large pigment clump in the upper part. At first the lens surface was quite free of pigment. Later on it became more and more covered by dense pigment dust, and many pigment clumps were seen, especially axially. In addition to the white pupillary seam in Fig. 323, there is peripherally a second light stripe. This latter is not in the iris, it is in the cornea and represents an opacification along the limbus.

In the limbus there is (physiologic) pigment, extending into the conjunctiva. It is also present in the other unexposed eye.

Figs. 284 and 325 to 328. Ring of depigmentation in a one year old gray brown rabbit, occurring 8 weeks after the exposure.

Regarding the cataract see Fig. 284. During the period of depigmentation there was much pigment dust, clumps and in heaps, visible in the aqueous, most of it had gravitated to the bottom of the anterior chamber. This pigment disappeared later on. A few days after the depigmentation, which at first exposed a delicate pupillary seam, a decided increasing *proliferation* of pigment occurred.

Large round pigment clumps at first nasal and temporally, then also below, presented themselves under the pupillary border. (Compare Fig. 282.)

Figs. 325 and 326 show the pigment changes $1\frac{1}{2}$ weeks, Fig. 327 $2\frac{1}{2}$ weeks after the exposure.

Fig. 328 presents the stage of pigment proliferation four weeks after the exposure. In this case the iris presented more decided changes in its upper zone. In this area it is atrophic and there are no pigment proliferations. In this direction the lens also presents greater changes, being contracted on account of absorption of the zonula.

Here also a circumscribed corneal opacification was observed, and the pterygium-like development seen in the illustration resulted. (Fig. 282.)

G.
EXAMINATION OF THE VITREOUS

1. THE NORMAL VITREOUS

In aphakia with ordinary focal illumination, especially if sunlight or an arc-light be used, it is possible to see the framework of the vitreous. It has a wavy tunic-like form and often extends forward into the anterior chamber.

I have at times seen this scaffold-like structure before the introduction of the slitlamp. *Gullstrand*[1]) was the first to observe the framework of the vitreous in the non-aphakic eye, with the slitlamp. These observations were confirmed by *Ergellet*[102]) and the author[49]), who have reported and described changes in this structure. Other observations regarding the forms and types and of the pathology of the framework of the vitreous have been reported by *L. Koeppe*[104]) and *J. Koby*[105]).

In spite of these reports we are still at the threshold of this knowledge, and some of the reported findings have not been fully confirmed. Pathologic anatomical findings and clinical observations must go hand in hand to register progress.

The normal vitreous supporting structure presents many variations and the distinctness with which it may be seen varies in different individuals. In some persons it may be easily studied with the Nernst or nitrogen lamp, while in other this type of illumination presents the vitreous as being quite "optically empty".

In general we may say, in agreement with the anatomic findings of *E. Fuchs*, that the framework has a pronounced *membranous lamellar* form. A fibrillar structure of the membranes may often be observed. The latter change is quite distinct in pathologic cases. Certain fibrils may undergo absorption. We must not confound these fibrillae with the luminous threads which are normally found quite frequently in the anterior portion of the vitreous, especially in old age.

In certain normal cases we receive the impression of a distinct *fibre structure* of the scaffolding. However, on increasing the intensity of illumination (micro-arc-lamp), we usually find it presents a lamellar form.

The specific weight of the framework is but little in excess of that of the vitreous fluid, so that it sways pendulum-like and is thrown into folds like a suspended cloth, on oscillation of the eyeball. If for instance, it is thrown upward on motion of the eyeball, it immediately returns to its original position.

As the framework is attached to the pars caeca, and the latter is connected anteriorly, as has been anatomically shown by *Salzmann*[16]), it is evident, why in the upright position of the body, the folds in the membranes of the framework assume a vertical direction.

The framework as a rule does not extend anteriorly to the lens surface. At this place there is a relatively "optically empty" space, filled with tissue fluids and aqueous, described by *Erggelet*[102]), and by the author[49]).

Koeppe has confirmed this observation with the slitlamp and *E. Fuchs*[106]) has recently described it anatomically.

Quite often, for instance in the case of Fig. 330, the vitreous is separated from the retro-lental space by a characteristically folded membrane, the much discussed membrana hyaloidea.

Especially *E. Fuchs*[106]) has anatomically shown that a *pseudo-membrane* of this kind, brought about by the apposition of lamellae, really exists. I have often observed it with the slitlamp. Fig. 330 shows it as a thin layer thrown into folds.

In Figs. 342 and 344, according to my judgment, it probably gives rise to a formation of strands. Similar curtain-like membranes may at times be seen at various deeper intervals, within the vitreous. *Gullstrand*[128]) had previously noted their approximate position.

At the time I changed my method of observation, substituting for the Nernst-lamp a *micro-arc-lamp*, creating a *micro-arc-slitlamp*, which latter was constructed by *E. Zeiss**, I was astonished at the increased amount of *new detail* visible in the vitreous, in areas which formerly seemed void of structure.

Some of the latter spaces now presented a finely fibrous and delicately meshed network of structure. The coarse lamellae which formerly were hardly visible now presented a high degree of opacification. On the posterior lens surface embryonic vessel remnants in great numbers, which were never before observed made their appearance.

The remnants of the hyaloid artery present their most minute detail in distinct outline. Even under the highest magnification the light is of a sufficient intensity to show all of these structures. With the aid of this new source of light in some cases the just mentioned "optically empty" retrolental space was practically absent, or existed as a dark space only in a very limited area. In the light of the micro-arc-slitlamp as far as I have been able to observe there is *no vitreous which is free of framework*, not even in the axial area.

In cases of average corneal curvature we may bring the vitreous into direct view with the bundle of light of the micro-arc-lamp to a depth of about 5 mm. Regarding the *form of the framework* there have been classified a series of more or less typic forms, especially by *Koeppe*[104]). We will not enter into detail regarding these, as their observation is somewhat dependent on the variety of light used, and further research is necessary.

Common types of framework as we have found them with the micro-arc-lamp are shown in Figs. 348 to 351. Some of these were absolutely invisible with the Nernstlamp illumination.

Note the *stepform layers* which illustrate the concentric structure of the framework of the vitreous (Figs. 339 and 351), the retrolental space (Figs. 344, 346), and the *vertically folded membrane with its dark cross striping* (Figs. 330, 346). These cross stripes give a characteristic design to the structure of the framework, and as I have frequently convinced myself, are also due to *folds*, that is, they represent a *cross folding of the membrane*. Especially luminous in comparison to the membranous framework are the fibre-like structures within the anterior area, as shown in Figs. 331 and 335.

Fig. 329. The determination of depth in the vitreous.

It is important, to localize areas in the vicinity of the *posterior lens capsule*. It is not always easy to definitely determine whether an object is on the posterior lens capsule or just behind it. One must use the same principles that govern the determination of depth in the cornea and lens, that is by focussing the sharp image of the bundle of light in such a manner onto the object that we may obtain the image of a "sagittal optical section". The slit through which the nitrogen lamp image passes must be narrowed so that the focal beam will not exceed a width of 0,1 mm.

---

* I am especially indebted to *Prof. Henker* for his assistance.

Place the sharply outlined posterior capsule strip L into exact focus (Fig. 331). This illustration shows the posterior lens capsule and the adjoining vitreous of the right eye of a young man. The source of light is temporal (shown by the arrow). As has been explained before, for the purpose of localization, let the bundle of light wander from side to side on the posterior capsule strip L. Whatever object appears and disappears within the line fh is in the area of the posterior capsule.

In Fig. 329 (horizontal section), C represents the capsule, L the illuminated capsule strip, and F the bundle of light. ' The double arrow shows the direction of observation. The points p and p' are in a sagittal direction of one another. Let us assume the bundle F is displaced parallel to itself and is at first in C. If we now move the bundle parallel to itself nasalward from C (in the direction of the arrow), point p will first appear, whereby it is situated *outside* of the luminous capsule strip. It is thereby localized as being behind the capsule. By further advancing in a nasalward direction, point p will also appear, and at the moment of its appearance it is on the *border* of the illuminated capsule strip (that is in the line fh, Fig. 331). For the observer, point p' is also in the same direction, its position *behind* point p, has been proven by the movement of the bundle of light, because during this movement it appeared *outside* of the capsular strip. In this manner it is possible to determine the position of various points in their relation to the lens capsule. Binocular vision is of great aid in this work.

These observations demand a distinct sharp uniform luminosity of the cross-section of the bundle of light.

Fig. 330. Normal framework of the vitreous of a boy 10 years old. The so-called membrana hyaloidea.

Nernstlight. Oc. 2, Obj. a2. Posterior to the lens and separated from it by a distinctly "optically empty" retrolental space, one is confronted by the very thin folded luminous reflecting membrane which is shown in Fig. 330. We frequently find this membrane especially distinct in normal eyes in youth. Axially it approaches the posterior lens zone, peripherally it is somewhat more distant, conforming in curvature somewhat to the posterior surface of the lens.

By "palpating" this membrane with the bundle of light it presents *extreme thinness* as a characteristic. It floats and waves on bulbar motion and a flattening of the folds occurs at times. By adapting the illumination I could prove in this and in other cases the existence of *real folds*, vertical and *diagonal* ones, the latter especially characteristic (see illustration).

These give origin to a featherlike design (Figs. 333, 346), just as can be brought about in a hanging robe which is supported from below or shirred. In view of the slight difference in the specific weight of the membrane compared to the surrounding fluid vitreous the effect of gravitation is but little in evidence.

In Fig. 330 this anterior folded limiting membrane only is shown, not the vitreous behind it. Just posterior to the membrane there is an "optically empty" space, after which further membranes are visible. Figs. 342 and 344 show folds due to traction, the substratum of which, in view of its position and extremly thin character, I must presume to be the "limiting membrane" of the vitreous. Whether we designate this layer as a "membrane", or "pseudo-membrane", or otherwise is only a question of expression. The fact is that it is proven present clinically as well as anatomically.

Fig. 325—335.

325

326

327

328

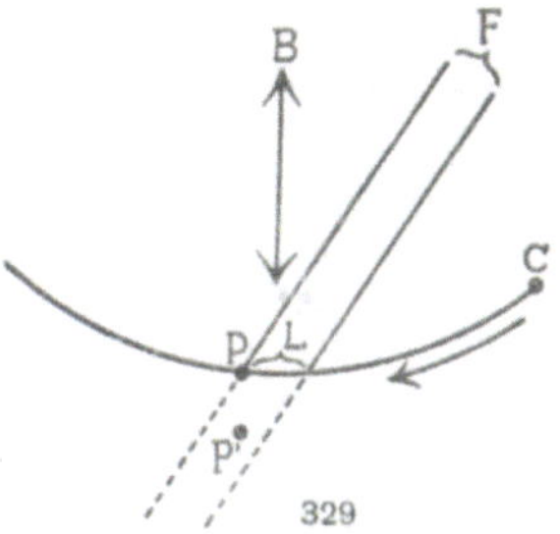

329

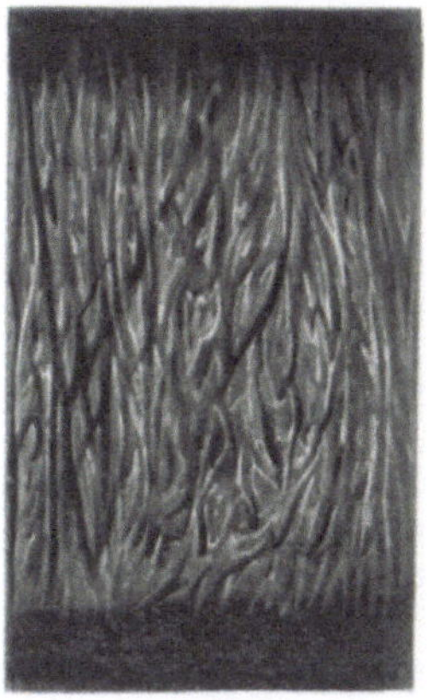

330

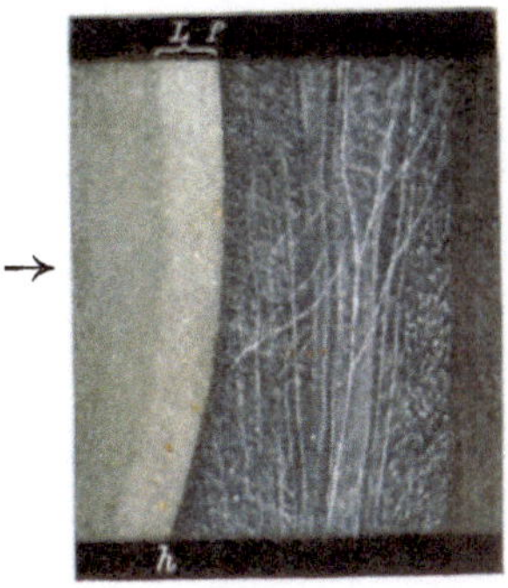

331

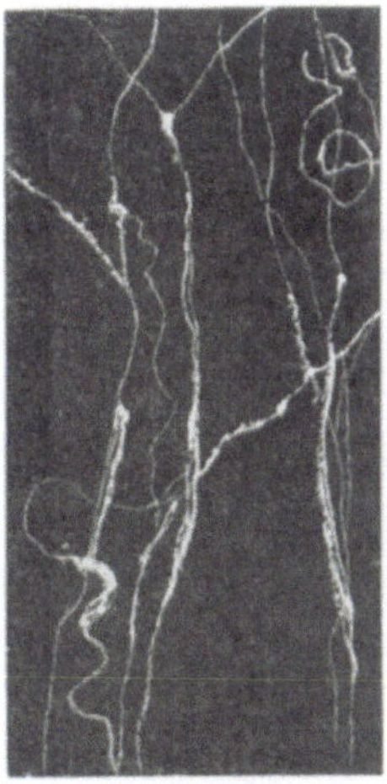

332a

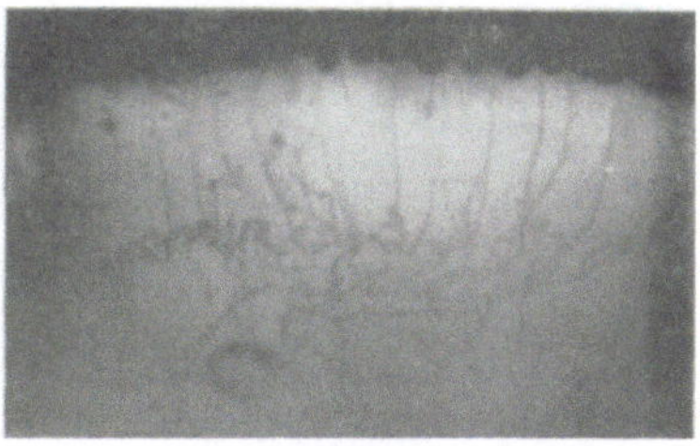

332b

333

334

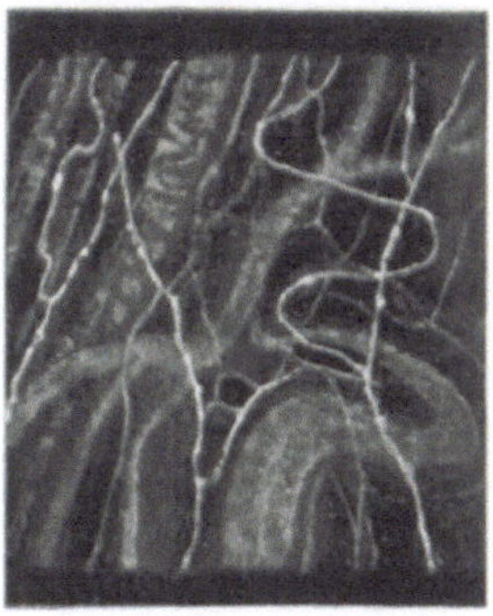

335

Verlag von Julius Springer, Berlin.

Fig. 331. Normal framework of the vitreous of a man 36 years of age in an optical meridional section.

Oc. 2, Obj. a2. Nernst slitlamp. Mechanic S. S. The posterior lens capsule axially presents golden-brown deposits (evidently pigmented), measuring 0,04 mm, and in the anterior part of the framework two or three white shining dots, of a type seen in pathologic processes. The framework is normal in structure. Vertical lamellae and fibres show horizontal and diagonal kinks and folds.

As the other eye had sustained an infected perforating wound, we at first suspected the vitreous dots shown in the illustration of being symptoms of a threatened sympathetic ophthalmia. No further development of dots occurred, and subsequent observations confirmed that they were probably congenital deposits, such as are frequently found present in small numbers. They may possibly be wander cells such as have been anatomically proven present in the vitreous *(Koeppe)*. Small individual pigment clumps are also found in normal eyes on the posterior lens capsule and in the vitreous.

Fig. 332a and b. Threadlike structures in the anterior, especially the peripheral, vitreous areas. (Remnants of the vasa hyaloidea propria?) Compare Fig. 164 and 165.

Arising near the origin of the arteria hyaloidea, that is from the arteria centralis retinae in the early embryonic months, there are present a number of small vessels which extend toward the peripheral vitreous areas, which *Kölliker*[67]) has designated the vasa hyaloidea propria. (As was found by *Schultze*[68]) they where obliterated in human eyes in the fifth month.) These vessels extend anteriorly in the form of extensive arcade formed loops into the peripheral zones of the tunica vasculosa lentis. They carry arterial blood, which flows in an anterior direction.

According to *O. Schultze*[68]) there are in all about 20 to 30 branches which originally extended in an anterior direction quite adjacent to the retina.

Posterior to the ciliary processes they form delicate loops, with their convexity forward, and join the branches of the arteria hyaloidea which comes from the posterior lens surface.

They are however not always limited in their location to the surface of the vitreous, but as has been first described by *Kessler*[65]), they may be found in all layers of the vitreous in later stages of development. They form a densely branched extensive network of vessels. *H. Virchow*[66]) also demonstrated a dense irregular meshwork of vessels. According to *Schultze's* findings in two human fetuses, whose bulbar diameter was 3 mm, the extensively meshed network of vessels which arises in the manner as has been described is more fully developed on the temporal side. I have injected these vessels in three to five months fetuses (I am indebted to the obstetrical clinic of the University Prof. *Labhardt*, for these specimens) by the way of the carotids.

Fig. 332b presents a microphotograph of the retro-equatorial zone of a fetal lens, the tunica vasculosa of which I injected with *Berlin-blue* by way of the carotid artery (fetus length 26 cm). (Photographed by cutting a window into the sclera.) One can see the anastomosing network connecting the vasa hyaloidea propria, with the vessels of the tunica vasculosa posterior, behind the lens equator. Extended anteriorly we may follow the parallel vessels surrounding the equator, as they disappear behind the round prominences of the ciliary body.

With the slitlamp one may observe in most infants after birth, especially peripheralward directly behind the lens, threadlike anastomosing structures as are shown in great numbers in Fig. 332a. These create the impression that they are *vessel remnants* and according to their location in part may be remnants of the numerous arcade-formed anastomosing loops of the vasa hyaloidea propria. At times these vessel remnants are found exclusively within the retrolental area. At other times they extend over the whole of the area posterior to the lens.

I have also found these structures in rabbits, pigs, dogs, calves and cats.

Regarding the differential diagnosis between them and remnants of the arteria hyaloidea compare the text to Fig. 148—165.

Fig. 333. Numerous punctate deposits in the vitreous in an apparently normal eye. Miss L. G. age 12.

The patient was brought to me on Feb. 10. 1919 on account of myopia (0,75 D Javal 0,5 D).

Both fundi, also under redfree light illumination, were normal. Seen with the slitlamp the right vitreous was free of deposits, while the left presented the numerous white deposits shown in Fig. 333. There were hyaloid remnants of the usual spiral form. The patient never had suffered any disease of the eye nor reported an ocular trauma. A relatively low bilateral central visual acuity of $1/3$ with correction was found. Under focal illumination with the ophthalmoscope the vitreous presented no changes, while with the redfree light there were delicate dustlike opacities. We are probably dealing with a pathologic process. On Oct. 13. 1919 eight months later, I was surprised to find only a few, some luminous, remnants of deposits. Early in December they were the same. They were more distinct and numerous with the micro-arc-lamp than with the Nernstlamp. There also were delicate dustlike opacities, in other respects the eyes were normal. I have found similar dense stationary vitreous opacities in the healthy eyes of two patients who had sustained a perforation of their other eyes. (B. A. aged 21, Vis = 1, and Miss R. W. aged 10.) There were no signs of sympathetic inflammation. Since then I have further observed some normal cases with punctate deposits in the vitreous of both eyes.

Fig. 334. The senile framework of the vitreous. Mrs. B. age 80.

Oc. 2, Obj. a2. Observation with the micro-arc-slitlamp.

The senile framework is at times characterized by an increased luminosity. In this case there are present the typic tunic-like vertical folds, which are especially bright. Anterior to this lamellar structure there is a network of threads and fibres, which no doubt is composed of vessel remnants. In this case the threads and fibres present a delicate white granular frosted appearance, so that they appeared as if covered with sugar and small wartlike excrescences. The substance between the membrane and threads also present a dense dustlike infiltration.

These changes, which have first been described in a similar manner by *Koeppe*[104]), and in our clinic by *Koby*[105]) are *not always* found in the senile vitreous by observing with the micro-arc-slitlamp.

2. THE PATHOLOGIC VITREOUS

Aside from senile changes, which may be classified as normal, the pathologic vitreous is especially characterized by *punctate deposits.*

(The first accurate observations were made by *Erggelet*[102]).)

These must not be confused with the punctate and knob-like dense formations which are found in small numbers on embryonic vessel remnants, or the sparce dots which have been histologically identified as being wander cells.

We are rather dealing with large numbers of deposits of a definite kind which are mostly found attached to the meshes of the framework of the vitreous. The nitrogen and especially the micro-arc-slitlamp expose the exact morphological relation of their structure to the framework.

They are either punctate to round deposits or platelets strewn into the framework (for instance in amotio retinae and retinitis pigmentosa), and attached to it, or they *present parts of it.* The latter is often the case in iridocyclitis. In this condition they are often *starformed* in shape, by virtue of the delicate framework threads which radiate from them into various directions, combining with other similar structures. They remind one of the form of the star figures, which are seen on the altered pigment deposited on the anterior lens capsule. The latter however are flat compared to the star-shaped deposits found in the vitreous.

The morphological differentiation of the deposits in the vitreous was formerly difficult because with the Nernstlight the detail, such as the threadlike proliferations as well as the more minute deposits and parts of the framework itself, were invisible. The nitrogen lamp is an improvement, but it remained for the arclight to first expose all of these delicate elements of threads, fibres and meshwork. The deposits are mostly composed of cellular elements. Hemorrhages produce uniform dustlike deposits, while the exudates arising from inflammation are more coarse in their nature, at least after existing for some time.

The red colour of the blood is better seen macroscopically than microscopically. That individual red cells never appear red, has been mentioned in the introduction of this atlas. They are seen as white to yellowish luminous dots. One must not neglect to inspect the zone of the luminous bundle macroscopically when examining the vitreous. It must also be emphasized that the use of the slitlamp does not eliminate the necessity of examining the vitreous under focal illumination with the ophthalmoscope. One important reason is that the slitlamp allows us to inspect the anterior vitreous only.

In certain diseases, in amotio retinae, and often following contusio bulbi, there is noted a *red* to *brownish red colouring* of the *deposits.* This is evidently due to uveal or retinal pigment. Bloodpigments may also participate in this colouration. It would however be going to far to attempt to differentiate the various blood derivatives by slitlamp observation.

When using a higher magnification one must avoid confusion by the aberration and diffraction of light. According to the source and variety of illumination and the chosen magnification, deposits may appear white or with a slight tinting, which may give origin to errors in diagnosis. The apparent *form* and *size* of deposits is also subjected to variations, according to conditions of refraction and reflection, and deceptions of a high degree may occur.

With the micro-arc-lamp I was able to observe delicate deposits which were invisible under other forms of illumination. In addition to deposits we also find pathologic changes of the *framework* of the *vitreous*. The framework normally presents many different forms and these are greatly increased in variety in pathologic conditions. The framework in one and the same vitreous often presents various forms. Coarse strands may alternate with cotton-wool-like wavy masses or with bundles of fibres and bands.

One may often observe localized increases in density, an apparent condensation, compared to the irregular folds of the framework. In high myopia a dissolution of the framework is often discernable. Large areas seem "optically empty" and on motion of the eyeball, balls of woolly or fibre-like framework appear. They float about more freely than in normal eyes.

That the vitreous opacities in iridocyclitis and degenerative myopia are mainly composed of cellular deposits onto the framework of the vitreous is quite evident from the illustration:

Fig. 335. Framework of the vitreous in myopia of 14 D. Miss O. F. age 45.

Oc. 2, Obj. a3, Nernstlight. The vitreous in large areas appears "optically empty". Large ball-like masses of framework composed of membranes and threads appear in rapid motion on bulbar movement.

They do not gravitate, but usually return to their original location, so that we must presume that they are attached to a definite vitreous supporting structure. (The "Glaskörperbasis" of *Salzmann*.) Anteriorly we may observe the luminous threads, which we presume are remnants of the vasa hyaloidea propria. Behind these the grayish lamellae, which in parts show the characteristic cross foldings are seen. There is present a central choroideal atrophy and metamorphopsia. By focal light with the ophthalmoscope there are large motile vitreous opacities. In the changes of myopia the framework of the vitreous is often found greatly displaced and mixed.

Fig. 336. Vitreous prolapsing into the anterior chamber. Mrs. P. age 64.

Low magnification. Seven weeks after complete luxation of the lens by contusio bulbi. In addition to the luxation of the lens, which was followed by a slightly increased tension, there are several superficial ruptures of the sphincter of the iris. The folds in the framework of the vitreous are vertical and extend forward into the anterior chamber (not visible in the illustration). They are densely covered by yellowish red dots (pigment debris). There is no pigment border visible on the pupillary edge. The pupil is dilated on account of the ruptures and because of tension.

Fig. 337. The area of prolapsing vitreous of the preceeding case under higher magnification.

Oc. 2, Obj. a2. A ball of cotton-like vitreous tissue is attached to a threadlike strand and covered by pigment clumps. The pigment is not detached but is adherent to the spiderweb-like strand seen in the illustration*.

---

* Herneas of the vitreous following cataract extraction are not rare. They were first observed by *Erggelet*[102]). I have never seen them unless there was a rupture of the posterior capsule or zonule. Recently I noticed a blood injected vitreous hernia following contusio bulbi, which hung over the lower pupillary border, into the aqueous. The lower iris *bulged forward*, there were a few ruptures of the sphincter, slight iridodonesis, while the lens seemed normal. The hernia was smaller, though still injected with blood, after two months. This case shows that hemorrhages which are attached to the framework of the vitreous resorb very much slower than such as are free in the aqueous. (Compare also the recent observations of *Hesse*.)

Fig. 336—345.

Tafel 36.

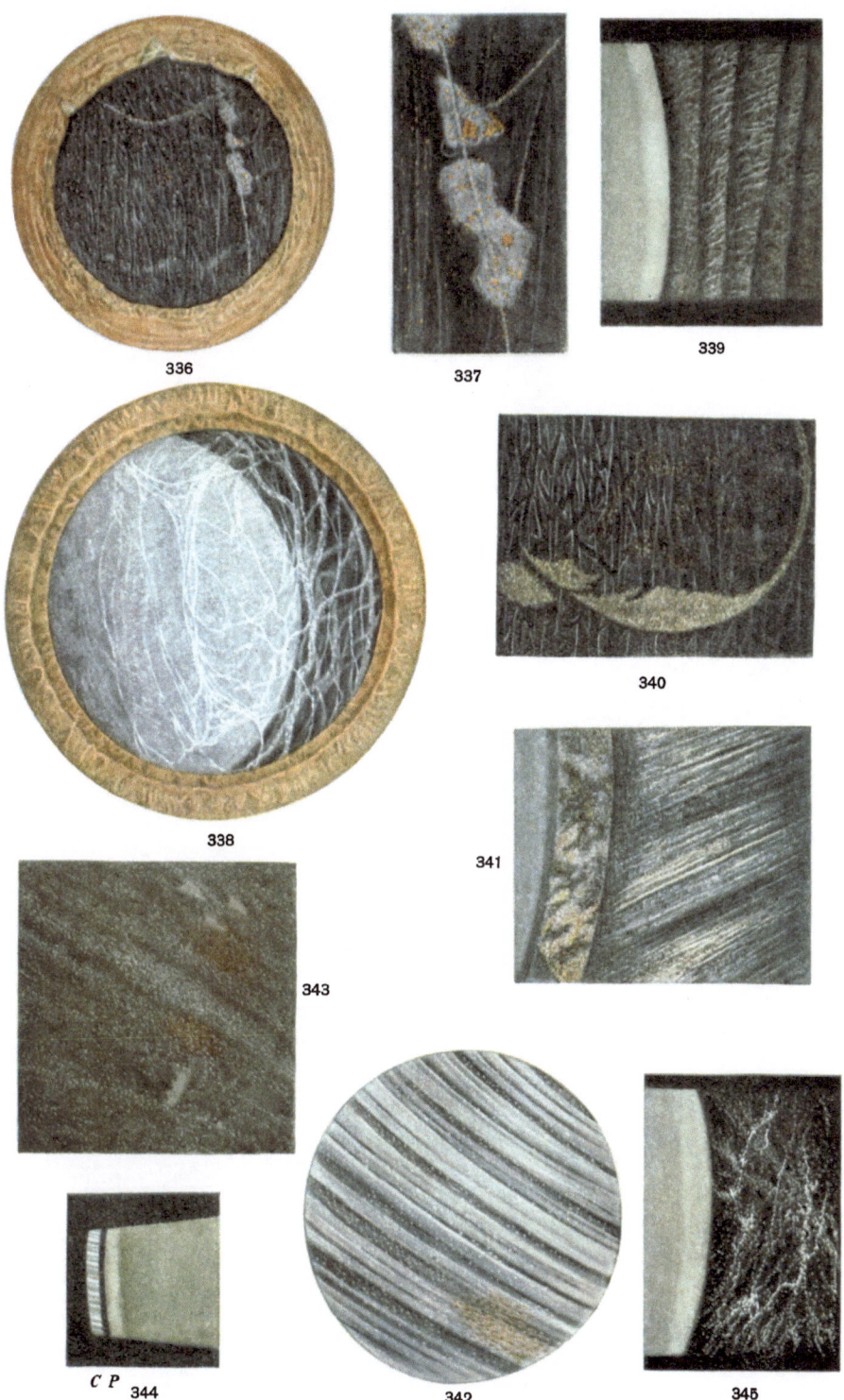

336

337

339

338

340

341

343

C P 344

342

345

Verlag von Julius Springer, Berlin.

Fig. 235—240.

Fig. 338. Subluxation of the lens with prolapse of the vitreous into the anterior chamber. Mr. S. Sch. age 80.

Sustained, one year ago, a subluxation of the lens by contusio bulbi, so that it is attached only to the temporal and upper temporal zonula, and on raising and lowering the head, appears and disappears, as if on a hinge, in the manner of an opening and closing door. The lens is opaque, the pupil dilated, tension normal and ophthalmoscopically the fundus is normal.

In the nasal area of the pupil, where the lens does not interfere, there are vitreous strands extending forward. They are somewhat more luminous in the illustration than when seen with the Nernstlamp. Vis R. = $^2/_{200}$.

Fig. 339. Hemorrhage into the vitreous. (Juvenile.)

Oc. 2, Obj. a2, Nernstlight. **Mr. A. S.** age 17, tall and anemic, suddenly lost his vision of the left eye. The ophthalmoscopic examination disclosed a dense hemorrhage into the vitreous. After its resorption several peripheral zones of retinitis became visible. There are similar retinal changes in the right eye. Note the step-like antero-posteriorly arranged normal area of folds, with deposits of delicate white to yellowish luminous dense dust, which in all probability are erythrocytes. The individual bodies are more distinct and some are luminous by using the micro-arc-lamp under increased (86 times) magnification. In this light they are pure white. The delicate dust which covers the framework of the vitreous in hemorrhages of this kind is not uniformly distributed, only being found present in certain individual areas which are distinctly separated from one another by dark intervals. (Areas free of supporting structure.)

That the blood occupies peculiar strand-like areas, especially following trauma, is shown in Fig. 341 and 343. Blood derivatives when observed under slitlamp illumination cannot always with certainty be differentiated from deposits of an inflammatory origin.

The latter may at times present a dust form type and be luminous in appearance.

Fig. 340. Hemorrhage into the vitreous and onto the posterior lens capsule.

Leutenant H. H. age 25. Contusio bulbi eight days ago. The right eye was hit by a pear.

The vitreous which was saturated with blood, cleared up rapidly. Today there is seen a delicate dustlike opacification of varied density in all of the vitreous, and a layer-like deposit of red blood corpuscles on the posterior lens capsule (temporal above). The deposit is in part ringformed, yellowish red, has a bronze-like lustre and reminds in its appearance of the "Vossius Ringtrübung". (Vossius' annular deposit on the anterior lens capsule following bulbar trauma.) Vision shortly after the injury $^{10}/_{200}$. Today $^6/_9$.

Fig. 341. Strandformed and striped hemorrhage into the vitreous following bulbar perforation.

Oc. 2, Obj. a3. **Mr. O. Sch.** age 16. While at work a knife hit his right eye, eight days ago. R. V. = $^{15}/_{200}$ tension reduced. Above the caruncle there is a perforation of the sclera measuring 2 mm situated 1 cm from the limbus.

Around the papilla there are a few and in the peripheral nasal fundus more extensive retinal hemorrhages. In the vitreous, especially anteriorly, most of the blood is found on *perfectly straight* diagonally horizontal strands, which apparently measure up to $^1/_2$ mm in diameter, and are vitreous framework under traction, saturated with blood (Fig. 341). On bulbar motion these strands but faintly participate in the usual gyrations of the framework, hence the definite impression that they are under traction.

During the resorption of the blood one could observe that the coarser strands are composed of more delicate fibres. All strands converge toward the site of perforation and connect the latter with the supporting structure on the opposite side of the vitreous. The balance of the framework of the vitreous presents glittering blood corpuscles and their derivatives. Some of them have been deposited on the posterior lens capsule. Eight days later the findings were the same, the hemorrhage was slightly reduced and the tension had returned to normal. (After a year these strands were still visible. The scleral scar showed traction from within. Vision = $^1/_4$.)

The development of these stripes was similar to those of Figs. 342 and 344.

In the case of Fig. 341, fourteen days after the injury, it was still impossible to locate the site of the perforation on account of the extensive hemorrhagic conjunctival oedema.

We would designate such strand formations in the vitreous following injury as a symptom of *perforation*. As they in part persist after years, they may point to a site of perforation, when all other diagnostic signs have become obliterated.

Figs. 342 to 344. Striping in the vitreous, in the vicinity of the posterior lens capsule following injury by an explosion.

The left eye of the 9 year old patient of Fig. 204. Four and one half years after an injury by a shot explosion. Amotio retinae downward. In the vitreous framework, within and between the fibres and lamellae, which appear loose and torn (not visible in the illustration), there are numerous punctate deposits.

Nasalward, almost directly posterior to the lens capsule, are the gray-white, sharply demarkated, tensely drawn stripes, with their ends slightly curved as shown in Figs. 342 to 344. Four years ago it was noted in the case history that there were deposits in the form of gray horizontal stripes on the posterior lens capsule.

There now are pigment and numerous white dots, as well as a large number of other dense deposit areas *on the posterior lens capsule.*

Fig. 342 shows a pigment area of this kind presenting dots arranged in a lineal manner in the direction of the stripes. Further temporalward the stripes are less luminous as they approach the capsule and become lost on its surface. *They are continuous with the horizontal lineal direction of the punctate dots on the posterior lens capsule,* shown in Fig. 343. These dots in places form brownish red groups, the majority however are white. Note the four indistinct gray dots which in all probability are flat deposits on the capsule.

Fig. 344 presents the slitlamp bundle of light as it passes through the nasal area of the posterior lens surface under a 10 times magnification.

Note the yellowish superficial surface strip *P* (other stripes presenting layers of lamellar separation are not shown), then the succeeding small *dark interval*, which is the retrolental space and following this the horizontally striped bluish band, which latter is an area of Fig. 342 under reduced magnification.

There are no opacities in the lens.

To what anatomical substratum does this horizontally striped surface belong?

From the history it seems that the left vitreous contained *blood coagula* for a period of months. (Later on an extensive retinal separation occurred.)

E. Fuchs[106]) has recently published the results of an anatomical investigation of the vitreous, in which he reported that frequently in the case of hemorrhages, the anterior vitreous border showed a thickening in the areas which separate the lamellae of the corpus vitreum from the retrolental space, and which may be designated as a form of membrana hyaloidea. Normally there is also present a membranous increase in density at this place, which surrounds the vitreous.

It may be assumed that in this case this pseudo-membrane (most anterior vitreous membrane), which is inserted into the area between the ora serrata and the ciliary processes was *tugged at and pulled* which trauma caused it to become opaque. This tugging also created the folds of traction of the opaque membrane, so that the condition as presented in Figs. 342 and 344 was brought about.

The vitreous strands dispose one to conclude that a *scleral perforation* had occurred, which at the time had been overlooked.

Traction on the vitreous framework may also explain the lineal arrangement of the *opaque dots* as are shown in Figs. 342 and 344.

Fig. 345. Framework of the vitreous, six months after a traumatic intraocular hemorrhage.

Oc. 2, Obj. a2. Master Bl. age 15, sustained a hemophthalmos right by being kicked by a horse, six months ago. The vitreous which was originally densely filled with blood, has now practically cleared up. In the posterior and medial areas there are fibrinous connective tissue masses firmly attached to the retina and vitreous. The framework of the vitreous is netlike, and is filled with dense delicate white platelets and dots. I have found similar conditions following all hemorrhages into the vitreous. In addition there were some vividly red dots, which are probably pigment. On the posterior lens capsule a few white dots may be seen.

Fig. 346. White and reddish punctate deposits in transcient sympathetic ophthalmia.

Oc. 2, Obj. a2. Nernstlamp. On and between the membranes of the framework of the vitreous there are many luminous reddish dots and platelets. This is the case of transcient sympathetic ophthalmia described on page 33 occurring in a healthy strong 10 year old boy of good family stock.

Oc. 2, Obj. a2. *Nine months later the vitreous was free of all deposits* (by Nernstlamp illumination!). A very few individual brownish-red bodies were seen deposited on the posterior lens capsule in this case, many more were found within the areas between the folds of the framework. That they are fixed and not in suspension, as has been incorrectly stated, is shown in this and many other similar cases of choroiditis, iridocyclitis and amotio retinae. After the framework has been in motion they are again found in the same exact relative position. The micro-arc-lamp shows the fine threadlike fibres onto which they are firmly attached. That these dots are most numerous in lower areas of the framework may be due to their weight. In Fig. 346 one may observe the "optically empty" retrolental space which has been described above. It is somewhat wider in the periphery than axially.

Fig. 347. Fibre-like meshwork of the vitreous with white and reddish-white punctate deposits. Mr. A. A. age 62.

Oc. 2, Obj. a2. Contusio bulbi two years ago by a tennis ball. In the other uninjured eye the framework of the vitreous is hardly visible by the Nernstlamp. There are apparently no dot-like deposits. At the time when the illustration was made the patient was under treatment on account of a slight veiling of both maculæ (indistinct white foci), which at first was etiologically considered due to myopia and arterio-sclerosis. Eight weeks later a very chronic iridocyclitis with tubercle formation on the pupillary border and a simultaneous tension developed.

It therefore remained undecided in this case whether the punctate deposits in the vitreous were a result of the iridocyclitis or due to the injury preceding it. I have often observed, as in this case, that a vitreous framework, which had been invisible or just faintly discernible, *became quite evident and luminous by the deposit of cellular elements.*

This was demonstrated in a most instructive manner by the second (right) eye in this case. At first with a dilated pupil and under careful scrutiny with the Nernst-slitlamp there was no framework visible axially and just faint traces of folds peripherally. Today, six weeks after an attack of bilateral iridocyclitis, the framework shows with an intense luminosity and distinctness. The cell-like deposits no doubt are the cause of this, however other still unknown changes may be contributing factors. The increased distinction with which the framework became visible presents the *incipiency of vitreous clouding.*

Step by step in this case, as in others, I have been able to follow the development of ragged connected opacities. They are but *parts of the framework of the vitreous which are covered by cellular fibrinous elements.*

In a similar manner hemorrhages may under certain circumstances give origin to combined circumscribed opacities (compare Fig. 345).

The cellular formations one may see under focal light with the ophthalmoscope as fine dots, socalled vitreous dust are especially distinct when seen with the redfree light.

Succeeding investigations of this case with the micro-arc-slitlamp demonstrated the type of the deposits to be starshaped. This makes it probable that the conglomerations in the vitreous of this case are iridocyclitic in nature.

Fig. 348, Punctate deposits in the vitreous in chronic iridocyclitis. Very likely tubercular in nature.

Oc. 4, Obj. a2. Beginning cataracta complicata. Observations by the micro-arc-slitlamp. Mrs. B. age 35, had for years suffered a chronic iridocyclitis with vitreous opacities and precipitates on the posterior corneal surface. There is incipient cataracta complicata (shown in Fig. 348, below) and a vivid iridescence of the posterior lens reflecting zone. The framework of the vitreous on account of dust and punctate deposits has become more dense and opaque during the past two years, and the lamellae and fibres have been changed into distinctive opacities, which here and there seem to surround more luminous areas (showing darker in Fig. 348).

The deposits are of a *star-shaped* type, that is the white foci, which no doubt are cell conglomerations, present fine delicate radiating framework threads.

This type is characteristic of iridocyclitis.

Fig. 349. The vitreous in the degeneration accompanying heterochromia.

Oc. 2, Obj. a2. Assistant C. M. age 19, RV = $^6/_{24}$, Hm 1,5, LV = 1 without glass.

The right eye presents colourless starformed, also some dustlike precipitates on the posterior corneal surface. The right iris is of a light grayish-blue, the left a brownish-gray.

Fig. **346—356**.   Tafel **37**.

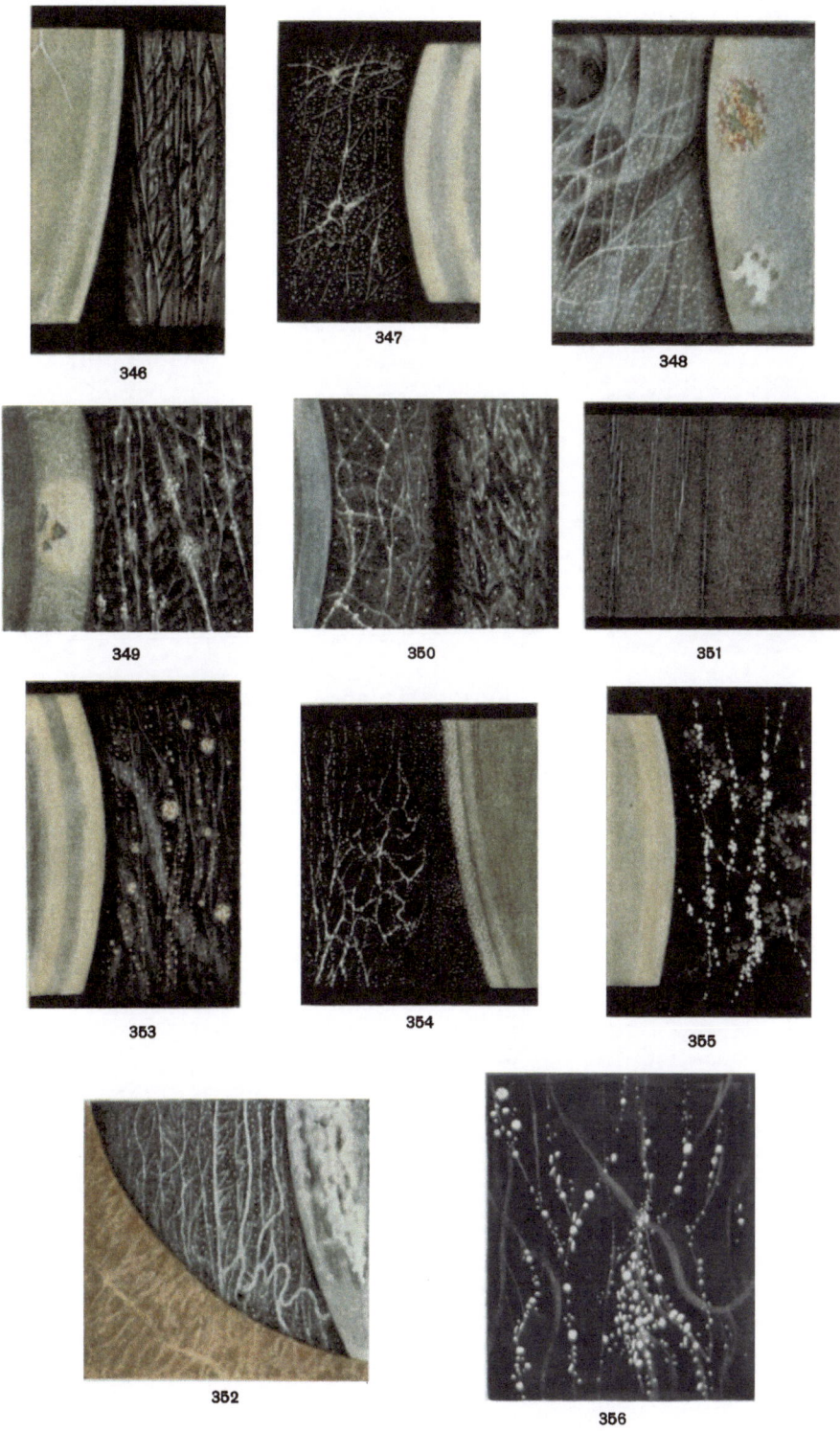

The right pupillary pigment border and retinal iris layer, as far as is discernible appears normal. The vitreous shows white to whitish-yellow flakes of varied sizes, which in certain areas are consolidated into groups and masses (Fig. 349). These are smooth and smaller than seen in synchysis, but similar without branching.

The framework of the vitreous is irregular, striped and ragged, in certain areas the membrane seems more dense and as if rolled into a ball.

The retina in the red-free light presents irregular preretinal reflex lines in the macular area. The yellow colouration of the latter is indistinct; there is evidently an exudate covering the membrana limitans interna.

The lens in the posterior cortex presents an increased reflection. Axially and in some peripheral areas the slitlamp discloses maplike iridescent subcapsular foci in the posterior reflecting zone.

On the posterior capsule one may observe, especially up and inward, stripe and comma-formed and punctate white deposits in great numbers. They remind of the threads and dots in corneal deposits.

This case of heterochromia was incidentally discovered when the other (left) eye, 5 days previously had sustained a rupture of the anterior lens capsule, with a prolapse of the lens substance by being hit with a stone.

Had one at this time overlooked the precipitates and opacities of the vitreous in the uninjured eye and discovered them subsequently, the erroneous diagnosis of "sympathetic ophthalmia" might have been easily made.

The right eye had been first observed when he was four years old, because of convergent strabismus. Three years later a tenotomy of the internus was performed, with the effect of bringing about a divergent strabismus.

The heterochromia at this time is not reported on the case record. It is not conspicuous at present, so that it was probably overlooked at the time.

Fig. 350. Punctate deposits in the vitreous in an old case of disseminated chor-oiditis, due to heredetary lues.

(Micro-arc-lamp 24 times lineal magnification.) Clerk Sch. age 23.

In his youth he suffered an attack of parenchymatous keratitis due to hereditary lues. Today the cornea still presents remnants of changes. The whole of the fundus is covered by foci of old chorio-retinitis of hereditary specific type. (In addition the patient presents a hereditary luetic involvement of the labyrinth.)

The vitreous presents disseminated delicate dustlike opacities with some larger white corpuscular deposits.

Fig. 351. Para-central choroiditis with dense delicate dustlike vitreous opacities.

Mr. R. S. age 24. Oc. 2, Obj. a3. Micro-arc-light. The delicate dustlike opacities are not visible by focal light with the ophthalmoscope, but one may observe a veil-like *circumscribed* opacity, not very easily set into motion.

The Nernst-light also fails to disclose the dense dustlike character of the opacity.

Within the dust-like area there is a delicate vertical striping, evidently composed of framework fibrillae. Note the *dust-free* dark intervals ("Saft-luecken", within which cellular elements may not adhere?) which are an additional proof that the vitreous is built up of concentric layers. The changes in the choroid are not very extensive. The right macula presents a few spotted foci below the fovea. In the left eye there is an irregular pigmentation of the macular area. The periphery of the fundus is normal. No deposits.

RV = $^6/_4$, LV = $^6/_5$. For the past three years the patient complained of flickering of light and difficulty in vision. When a child he had an attack of pleurisy. The Wasserman was negative.

Fig. 352. Punctate deposits in the vitreous in retinitis pigmentosa.

A. G. age 39. Micro-arc-lamp*. Illumination from the right, the bundle of light passing through the cataractous posterior cortex. Oc. 2, Obj. a2. Advanced retinitis pigmentosa, with cataracta complicata and a high degree of concentric contraction of the visual fields. The patient is somewhat deaf (central). Fig. 352 shows a lower zone of the vitreous. The pupil is dilated.

There are grayish white mostly vertical rope-like strands with a few white luminous punctate deposits within the framework. The deposits present no visible radiations, are white, small and of varied sizes. Aside of the strands, the framework is cotton-like and contains delicate dustlike deposits.

Fig. 353. The vitreous in retinal separation.

Oc. 4, Obj. a2, Nernst-light. Dr. K. age 62. He suffered a hereditary retinal separation in the right eye six months ago. (Hereditary, because two sisters of his mother lost their vision on account of bilateral myopic retinal separation.) Myopia of 6. D. in both eyes.

The vitreous allows us to study the changes characteristic of retinal separation.

Brown-red to red dots and clumps, evidently containing pigment or composed of it, are attached to the framework which latter is in many places dissolved into fibres, the whole being easily put into active motion. In addition there are delicate white punctate changes. In certain areas there are large globular structures, covered by punctate pigment deposits. A retrolental interval is not present.

(The posterior lens pole in this case showed a vivid iridescence and incipient cataracta complicata.)

The micro-arc-lamp fails to show star-shaped types of white or red deposits.

Fig. 354. Fibre-like network of vitreous framework containing dense white dots and white punctate deposits on the posterior lens capsule.

Oc. 2, Obj. a2. Miss B. R. age 8. The left eye is said to have been amblyopic since birth. The posterior vitreous presents dense floating membranes, which cover the nervehead. There are dustlike opacities. The retinal periphery is normal. There are also white dust-like deposits on the posterior lens capsule. (The cause of this evidently inflammatory change is unknown. There are no evidences of lues.)

Fig. 355. Synchysis scintillans under slitlamp illumination.

(The first slitlamp observations in synchysis were made by *Ergyelet*.[102])

Miss S. Sch. age 61, right eye. Bilateral senile cataract of the coronar type. Oc. 2, Obj. a2.

In synchysis scintillans the slitlamp presents, snow-white round or disc-shaped white particles of varied sizes up to 0,05 mm and larger deposited on the framework

---

* According to *Koeppe*, the deposits in the vitreous in all cases of retinitis pigmentosa are composed of pigment. We could not convince ourselves of this in any case. The dots in all of our cases were white to yellowish white, practically without exception.

of the vitreous. The latter is semi-fluid and seems degenerated. The deposits are arranged in rows or situated on layer-like surfaces. (The faintly visible conglomerations are not in exact focus.) The lenses of both eyes are slightly cataractous.

The synchysis is only present in the right eye which had always been somewhat amblyopic.

Fig. 356. Synchysis scintillans.

Mr. P. age 58, right eye. Oc. 4, Obj. a2. The spherical bodies in this case present a greater variation in size. There are diabetic changes visible in both retinae, while the synchysis is monolateral.

H.

APPENDIX

CONJUNCTIVA OF THE BULB AND LIMBUS

Slitlamp microscopy often discloses folds in the normal conjunctiva, with characteristic double reflex lines (Vogt[28]). (Compare microphotograph Fig. 90.) We could not diagnose the lymph vessels and sheaths first described by *Koeppe*[108]) with certainty, so we do not at present attempt their systematic presentation. (Compare text to Fig. 9.)

We have *very commonly* found anomalies of vessels, presenting peculiar varicose dilatations, characteristic in type. These dilatations repeatedly assume similar forms (Fig. 362 and 363). The first observations of these structures were by *P. Bajardi*[136])[137]), compare also the investigations of *J. Streiff*[135]).

We have presented pingueculae in two illustrations. The latter changes are greatly varied in form, as has been shown by the anatomical investigations of *E. Fuchs*[107]) *and others*.

Fig. 357 and 358. Senile pigmentation of the bulbar conjunctiva and episclera.

Not alone at the limbus, but also in the free bulbar conjunctiva we find pigment deposits in the stroma of the conjunctiva and episclera in old age, in which perivascular areas are free of involvement. Pigment areas of this kind are illustrated in Fig. 357.

It presents the right eye of teacher H. age 70 (case of Fig. 561. The light slate-coloured gray spot measures 7 mm. It is situated in the upper outer bulbar conjunctiva. Superficial vessels are movable over the spot. Evidently the pigment is mostly situated in the episclera. It is worthy of note that the pigment is often arranged in zones which are branched and apparently present a direction parallel to delicate vessels and capillaries.

Fig. 358 presents a pigment zone of this type under a high magnification (68 times). At the limbus I have observed that the vessels of the palisades following obstruction of the circulation have become modified and now present pigmented lines. We must therefore accept that senile "conjunctival naevi" are the results of hematogenous pigment deposits. The relation of this pigment to the vessels is of especial interest. The surroundings of the superficial conjunctival vessels (light red) and the superficial conjunctiva itself are free of pigment. The pigment is found in *a single layer of the episclera*, and in the latter is separated from the (bluish-red) vessels by *clear sheaths*. (Lymph sheaths?).

Non-pigmented stripes of this kind are also present without accompanying vessels. Note the white stripes which at times cross one another (Lymphvessels?).

Especially characteristic is the straight course and uniform width of these light stripes and sheaths. That the pigment is situated in *one* layer, is proven by the fact that it ends at the white sheaths with a sharply demarkated border, without extending anteriorly or posteriorly to them or to the vessels.

It therefore belongs to the same layer as the latter, which factor adds to the proof of its hematogenous origin.

Fig. **357—367.** Tafel **38.**

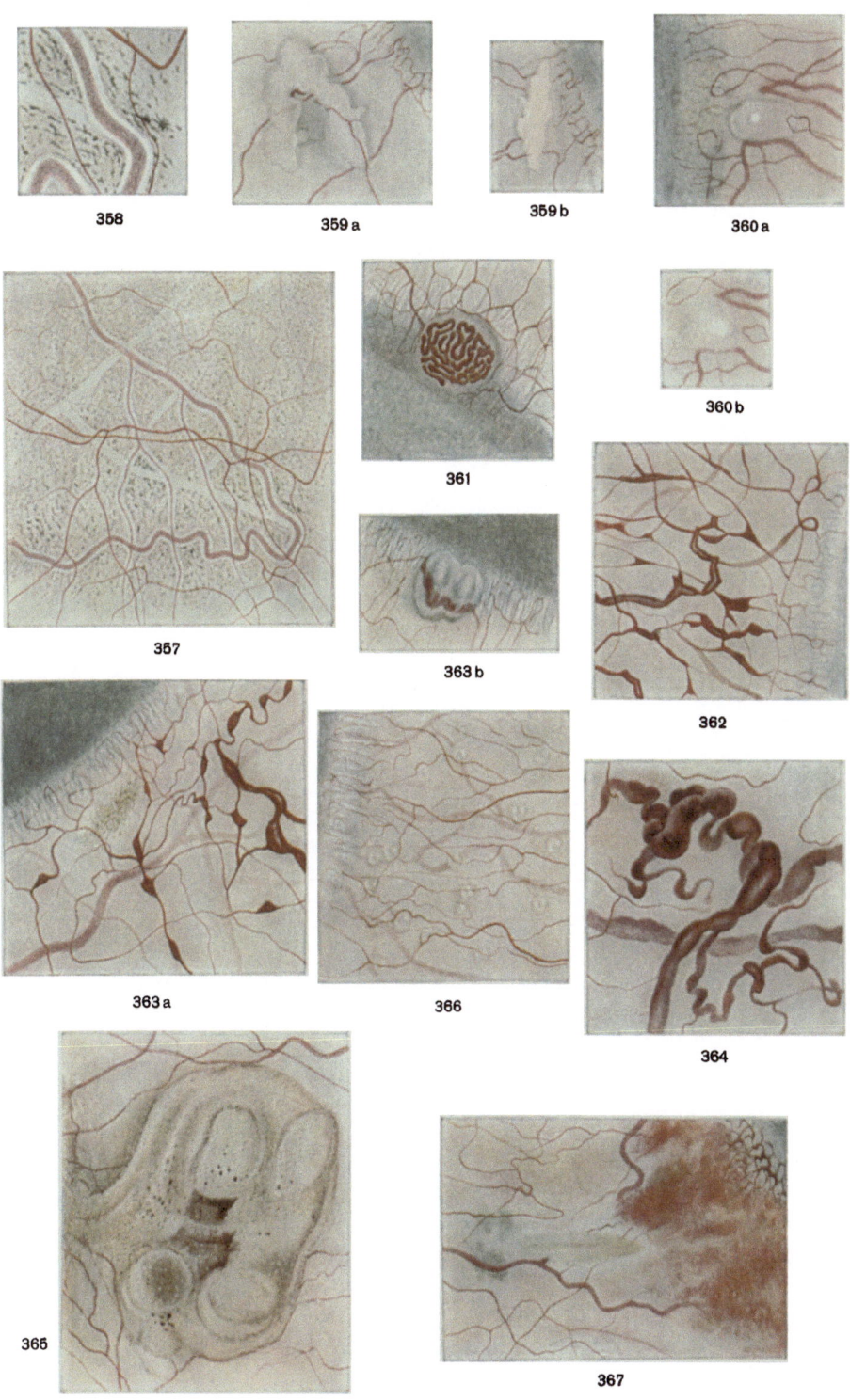

358

359 a

359 b

360 a

357

361

360 b

363 b

362

363 a

366

364

365

367

Fig. 357—367.

Fig. 359a and b. Nasal and temporal pingueculae. (Common types.) Mrs. M. age 70.

Oc. 2, Obj. a2. Fig. a = right eye temporal, Fig. b = left eye nasal. Fig. a shows a horse-shoe-shaped succulent grayish-white sausage like, sharply demarkated avascular mass, under a clear layer of the conjunctiva, near the temporal limbus.

The vessels pass in part over and under, some through this translucent structure. The widest area measures 0,3 mm. On the right nasal limbus a similar smaller change is seen. Fig. b presents the pinguecula of the left nasal limbus. It is displaced toward the limbus, is not sausage-like but presents an irregular yellowish white, non-succulent dry mass. It is so near to the cornea that it reminds of a beginning pterygium. Its width is 0,25 mm, its length 0,75 mm. The vessels are not visible through the opaque mass.

Fig. 360a and b. Phlycten-like vascularized vesicle of the pinguecular zone, with a calcareous area. Mrs. M. age 82.

Right eye. Nasal limbus. Oc. 2, Obj. a2. The white calcareous area (Fig. a) measures 0,3 by 0,5 mm. Fig. a presents it in a partial shadow as the light is directed toward the corneal border from the temporal side. The yellowish area of the corneal border represents the posterior reflecting zone.

Fig. b demonstrates the change in direct illumination. The calcareous area is somewhat obscured by the reflection of the conjunctiva which covers it.

The surroundings of the vesicle were vascularized for a period of weeks, so that a phlyctenule was simulated. Following currettement the injection subsided.

Fig. 361. Angioma of the limbus. Mr. H. age 70.

General arterio-sclerosis and bilateral hemorrhagic retinitis. On the right upper inner limbus there is situated a very small blood-red dot, which at first was taken to be a hemorrhage. The illustration shows the dot under a medium degree of magnification (Oc. 4, Obj. a2). It is a round convoluted sausage-like vessel. A vas efferens and afferens are not distinctly discernible. The convolution is flat, in one layer, and has a transparent covering. It is situated half on the sclera and the other half is on the conjunctiva of the cornea.

Three months after the making of the illustration the angioma remained unchanged. (Similar angiomata were found by *Koeppe*[139]).

Fig. 362. Telangiectatic and varicose changes of the conjunctival vessels. (Evidently only of veins.) Mr. R. M. age 28. (Bajardi[136])[137]).

Left eye, nasal bulbar conjunctiva. Oc. 2, Obj. a2. Especially in old age, sacklike, at times kinked and at the site of branchings presenting shrimp-skin-like forms, as well as perfectly circular dilatations of thin vessels may often be found. Ectasiae of this kind may be limited to a circumscribed area, especially within the interpalpebral zone. They may at times cause subconjunctival hemorrhages.

Fig. 363a and b. Varices of knot-like and spherical forms, in the superficial bulbar conjunctival veins. Fireman Sch. age 42.

Left eye. Temporal of and below the limbus. They are of a similar, though more knotted form, compared to the ones of the preceding case. Near the limbus there is a zone of pigment deposit.

Fig. b shows an isolated varix on the lower nasal limbus. This varix is within a double-walled cyst, which latter presumably is a perivascular lymph-cyst.

Fig. 364. Varices of the bulbar conjunctiva temporal area. Mr. J. L. age 69.

Oc. 2, Obj. a2. (The area is about 4 mm from the limbus.) The veins are unequally dilated. The convolutions are due to stasis.

The violet coloured vessels are more deeply situated. Both eyes present varices of this type in great numbers. In addition the patient has senile cataracts. At the time of the flap extraction, complications in healing or extensive hemorrhages did not occur. Varices of the conjunctiva are quite commonly found in eyes which are otherwise normal, especially in older individuals. (Compare *Coats*[117]).

Fig. 365. Filtration cicatrix of the conjunctiva following iridectomy for glaucoma.

Mrs. W. age 70, left eye. Two and one half years ago the left eye was iridectomized on the temporal side on account of glaucoma. The anterior chamber is shallow, the sagittal lens diameter is increased. There are a few scattered peripheral lens opacities. The tension now is normal. The visual fields are somewhat contracted concentrically. There is no excavation at the nerve head. Vision = $^1/_3$.

The cyst measures 4 mm vertically, and 3 mm in its horizontal diameter, and is composed of four or five chambers. (Two upper chambers have united within the past three months.) Cysts of this kind are best observed in indirect light, with the slitlamp. There are many pigment crumbs, which are washed out of the interior of the eyeball from time to time, attaching themselves to the partitions of the cyst and to its tissue framework. (Compare the observations of *Erggelet*[102]).

In this case the centre of the cyst presents a dark irregular rhomboidal brown tissue mass, on the border of which retinal pigment is visible. It is an incarcerated tag of iris. This incarcerated tag evidently caused the formation of the cyst and thereby brought about the end of the glaucomatous process.

Fig. 366. Conjunctival vesicle in scrofulous kerato-conjunctivitis.

Miss A. S. age 20. Oc. 2, Obj. a2. For the past few days she suffered a recurrent attack of superficial scrofulous keratitis.

Peripheral corneal areas show a delicate pannus, above there is a fresh circumscribed infiltration. The bulbar conjunctiva, especially in the interpalpebral zone shows many round clear elevations, some of which are under superficial vessels.

There is marked photophobia and epiphora.

Vesicle-like eruptions of this kind of a transcient nature are quite commonly seen in scrofulous kerato-conjunctivitis. (Miliary form of eczematous conjunctivitis, according to *Saemisch*[102].) The vesicles are best seen in the reflecting zone.

Fig. 367. Recent perforating wound of the conjunctiva.

Oc. 2, Obj. a2. A. B. age 40. Twenty four hours ago, while practicing with a rifle a foreign body entered his left eye. Nasalward of the limbus and at the latter a few conjunctival hemorrhages. Within this area a yellowish distinctly outlined lancette-shaped horizontal wound (Fig. 367).

Two vessels, one measuring 30 microns and another very much smaller one have been ruptured. Their separated ends are seen respectively in the upper and lower edge of the wound. To the left of the wound there is a dark area especially distinct in indirect light. This is the foreign body covered by translucent tissue.

Subconjunctival hemorrhage is an important symptom of scleral perforation. In this case the spicule of iron was imbedded within the sclera, and was removed by the magnet, through the original wound of entry.

LITERATURE

(The literature published during 1919 and 1920 could not be fully taken cognizance of)

1) Gullstrand, Allvar. Demonstration der Nernstspaltlampe. Vers. O. G. Heidelberg (1911), S. 374.
2) Vogt, Alfred. Zur Kenntnis der Alterskernvorderfläche der menschlichen Linse etc., Kl. M. f. A. **61**. 101. (1918.)
3) Koeppe, Leonhard. Klin. Beobachtungen mit der Nernstspaltlampe etc. Arch. f. O. G. **96**. 234. (1918.)
4) Henker, O. Ein Träger für die Gullstrandsche Nernstspaltlampe IV. 75. (1916.)
5) Vogt, Alfred. Der Altersstar, seine Heredität und seine Stellung zu exogener Krankheit und Senium. Z. f. A. **40**. 135. (1918.)
6) Helmholtz, H. Physiologische Optik I. Aufl. 1867.
7) Stokes, G. G. Pogg. Ann. **87**. 450. 1852. Erg. Bd. **4**. 177. 1854. Phil. Trans. 1852. S. 463 (vide Winkelmann, Hdb. d. Physik, 1906).
8) Spring, W. Bull. Acad. Belg. **37**. 174. (1899.)
9) Vogt, Alfred. Untersuchungen über die Blendungserythropsie der Aphakischen und Lichtexstinction durch die Cataract etc.. Arch. f. A. **78**. 93. (1914.)
10) Vogt, Alfred. Analytische Untersuchungen über die Fluoreszenz der menschlichen Linse und der Linse des Rindes. Kl. M. f. A. **51**. 129. (1913.)
11) Vogt, Alfred. Klin. u. experim. Untersuchungen über die Genese der Vossiusschen Ringtrübung. Z. f. A. **40**. 213. (1918); compare with more recent authors Triebenstein, Schürmann, Behmann.
12) Stähli, Jean. Zur Augenuntersuchung mit Nernstlicht. Beitr. z. A. **82**. 65. (1912.) Über Betauung comp. Koeppe, Arch. f. O. G. **96**. 199. (1918) und Vogt, Arch. f. O. G. 101. 123. (1920), further Vogt, Kl. M. f. A. Sept. 1920.
13) Stähli, Jean. Die Azoprojektionslampe (Halbwattlampe) der deutschen Auergesellschaft, ein Ersatz für Nernstlicht. Kl. M. f. A. **54**. 685. (1915.)
14) Vogt, Alfred. Der hintere Linsenchagrin bei Verwendung der Gullstrandschen Spaltlampe. Kl. M. f. A. **62**. 396. (1919.)
15) Vogt, Alfred. Die Sichtbarkeit des lebenden Hornhautendothels, ein Beitrag zur Methodik der Spaltlampenmikroskopie. Arch. f. O. G. **101**. 123. (1920.)
15a) Vogt, Alfred, Die Sichtbarkeit des lebenden Hornhautendothels. Ges. d. Schweiz. Augenärzte, Kl. M. f. A. **63**. 226. (1919.)
16) Salzmann, Maximilian. Anatomie und Histologie des menschlichen Auges. Leipzig—Wien 1912. S. 39.
17) Greeff, Richard. Pathologische Anatomie des Auges. Berlin-Hirschwald 1906. S. 117.
18) Hassal, A. The microscopic anatomy of the human body. London 1846.
19) Henle. Hdb. der systematischen Anatomie. 1866.
20) Hess, C. v. Pathol. und Ther. des Linsensystems, Hdb. v. Graefe-Saemisch, II. und III. Aufl. 1905 und 1911.
21) Vogt, Alfred. Das vordere Linsenbild bei Verwendung der Gullstrandschen Nernstspaltlampe etc. Kl. M. f. A. **59**. 514. (1917.)
22) Vogt, Alfred. Der menschliche Linsenchagrin und die Chagrinkugeln. Kl. M. f. A. **54**. 194. (1915.)
23) Schürmann, Josef. Weitere Untersuchungen über die Linsenchagrinierung usw. Z. f. A. **22**. 11. (1917) and Inaug.-Diss. Basel 1917.

24) Vogt, Alfred. Über Farbenschillern des vordern Rindenbildes der menschlichen Linse. Kl. M. f. A. **59.** 518. (1917.)

25) Purtscher, O. Ein interessantes Kennzeichen der Anwesenheit von Kupfer im Glaskörper. Z. f. A. März-April 1918, comp. Goldzieher, Hillmanns, zur Nedden, Ertl, Jeß, Pichler u. a.

26) Vogt, Alfred. Das Farbenschillern des hintern Linsenbildes. Kl. M. f. A. **62.** 582. (1919.)

27) Vogt, Alfred. Die Untersuchung der lebenden menschlichen Linse mit Gullstrandscher Spaltlampe etc. 41. Vers. O. G. Heidelberg (1818), S. 286.

28) Vogt, Alfred. Reflexlinien durch Faltung spiegelnder Grenzflächen im Bereiche von Cornea, Linsenkapsel und Netzhaut. Arch. f. O. G. **99.** 296. (1919.)

29) Koeppe, Leonhard. Klin. Beobachtungen mit der Nernstspaltlampe etc. Arch. f. O. G. **99.** 1. (1919.)

30) Koeppe, Leonhard. Klin. Beobachtungen mit der Nernstspaltlampe etc. Arch. f. O. G. **97.** 1. (1918.)

31) Virchow, Hans. Mikroskopische Anat. d. äußern Augenhaut und des Lidapparates. Hdb. v. Graefe-Saemisch, II. Aufl. 1910.

32) Stähli, Jean. Über den Fleischerschen Ring beim Keratokonus und eine neue typische Epithelpigmentation der normalen Cornea. Kl. M. f. A. **60.** 721. (1918.)

33) Meller, J. Über traumatische Hornhauttrübungen. Arch. f. O. G. **85.** 172. (1913), further: Über die posttraumatischen Ringtrübungen der Hornhaut. Kl. M. f. A. **59.** 62. (1917.)

34) Caspar, L. Subepitheliale Trübungsfiguren der Hornhaut nach Verletzung. Kl. M. f. A. **57.** 385. (1916.)

35) Pichler, Alexius. Die Casparsche Ringtrübung der Hornhaut. Z. f. A. **36.** 311.

36) Axenfeld, Th. (Herabsetzung der Sensibilität der Keratokonusspitze.) Discussion on the lecture of A. Siegrist, „Zur Atiologie des Keratokonus". 38. Vers. O. G. Heidelberg (1912), S. 193.

37) Koeppe, Leonhard. Arch. f. O. G. **93.** 215. (1917.)

38) Strebel und Steiger. Über Keratokonus. Kl. M. f. A. 51. 284. (1913), cp. further A. Vogt, Arch. f. O. G. **99.** 296. (1919) and A. Vogt, Zitat Nr. 76.

39) Axenfeld, Theodor. Zur Kenntnis der isolierten Dehiscenzen der Membrana Descemeti. Kl. M. f. A. **43.** 157. (1905.)

40) Haab, Otto. Das Glaukom und seine Behandlung. Sammlung zwangloser Abhandlungen, IV. Bd. 1902, comp. auch Protokollauszug in Arnold, Die Behandlung des infantilen Glaukoms etc., Beiträge zur Augenheilkunde 1891, S. 16. Further: Haabs Atlas der äußern Augenerkrankungen.

41) Stähli, Jean. Klinik, Anatomie und Entwicklungsmechanik der Haabschen Bändertrübungen im hydrophthalmischen Auge. Arch. f. A. **79.** 141. (1915.)

42) Heß, C. Klin. und exp. Studie über die Entstehung der streifenförmigen Hornhauttrübung nach Starextraktion. Arch. f. O. G. **38.** 1. (1892.) Further: The same author, Arch. f. A. **33.** 204. (1896.)

43) Schirmer, O. Über die Faltungstrübungen der Hornhaut. Arch. f. O. G. **42.** 1. (1896), cp. Treutler Z. f. A. III. 484. (1900.)

44) Dimmer, F. Eine besondere Art persistierender Hornhautveränderung (Faltenbildung) nach Keratitis parenchymatosa. Wiener Kl. Wo. 1905, S. 635 und Z. f. A. **13.** Ergänzungsheft 635. (1905.)

45) Heß, C. Beobachtungen über den Akkommodationsvorgang. Kl. M. f. A. **42.** 310. (1904), s. a. Graefe-Saemisch, Hdb. II. u. III. Aufl.

46) Vogt, Alfred. Klin. u. anat. Beitrag zur Kenntnis der Cataracta senilis etc. Arch. f. O. G. **88.** 329. (1914.) Further: Zur Frage der Cataractgenese etc. Kl. M. f. A. **62.** 111. (1918.)

47) Henle. Hdb. d. Anatomie, Bd. II, 1866, S. 682.

48) Barabaschew. Beitrag z. Anatomie der Linse. Arch. f. O. G. **38.** 1. (1892.)

49) Vogt, Alfred. Ein embryonaler Kern der menschlichen Linse. Korrespbl. f. Schweizer Ärzte 1917, Nr. 40. Further: Der Embryonalkern der menschlichen Linse und seine Beziehungen zum Alterskern. Kl. M. f. A. **59.** 452. (1917,)

50) Szily, A. v. Die Linse mit zweifachem Brennpunkt. Kl. M. f. A. **41.** 44. (1903), comp. Leop. Müller (1894), Demicheri-Tscherning (1895), Berlin (1898).

51) Heß, C. Über Linsenbildchen, die durch Spiegelung am Kerne der normalen Linse entstehen. Arch. f. A. **51.** 375. (1905.)

52) Vogt, Alfred. Die vordere axiale Embryonalcataract der menschlichen Linse. Z. f. A. **41.** 125. (1918.)

53) Meyer, G. Die Diskontinuitätsflächen der menschlichen Linse. Pflügers Arch. f. d. ges. Physiologie 1920, S. 178.

54) Rabl, C. Über den Bau und die Entwicklung der Linse. Leipzig 1900.

55) Vogt, Alfred. Die Spaltlampenmikroskopie des lebenden Auges. Münch. med. Wo Nr. 48, 1919, S. 1369.

56) Arnold, J. Beiträge zur Entwicklungsgeschichte des Auges. Heidelberg 1874.

57) Lüssi, Ulrich. Das Relief der menschlichen Linsenkernvorderfläche im Alter. Kl. M. f. A. **59.** 1. (1917.)

58) Vogt, A. und Lüssi, U. Weitere Untersuchungen über das Relief der menschlichen Linsenkernoberfläche. Arch. f. O. G. **100.** 157. (1919.)

59) Koeppe, Leonhard. Die Ursache der sog. genuinen Nachtblindheit. Münch. med. Wo. Nr. 15, 1918, S. 392. Derselbe, Z. f. A. **38.** 89. (1917.)

60) Brückner, A. Über Persistenz von Resten der Tunica vasculosa lentis. Arch. f. A. **56.** Ergänzungsheft 1907.

61) Koeppe, Leonhard. Klin. Beob. an der Nernstspaltlampe etc. Arch. f. O. G. **96.** 233. (1918), gleichzeitig Habilitationsschrift.

62) Vogt, Alfred. Der physiologische Rest der Art. hyaloidea der Linsenhinterkapsel und seine Orientierung zum embryonalen Linsennahtsystem. Arch. f. O. G. **100.** 328. (1919), comp. ibidem Bd. 101. Heft 2/3.

63) Bach und Seefelder. Atlas der Entwicklungsgeschichte des menschlichen Auges. Leipzig 1911.

64) Vogt, Alfred. Die Untersuchung der lebenden menschlichen Linse mittelst Spaltlampe usw., Vers. O. G. Heidelberg 1918.

65) Keßler. Zur Entwicklung des Auges der Wirbeltiere. Leipzig 1877.

66) Virchow, H. Glaskörpergefäße und gefäßhaltige Linsenkapsel bei tierischen Embryonen. Sitzungsber. d. physikal.-mediz. Ges. zu Würzburg 1879.

67) Kölliker. Lehrb. d. Entwicklungsgeschichte des Menschen und der höhern Säugetiere, II. Aufl. 1879.

68) Schultze, O. Zur Entwicklungsgeschichte des Gefäßsystems im Säugetierauge. Festschrift f. Kölliker. Leipzig 1892.

69) Vogt, Alfred. Beobachtungen an der Spaltlampe über eine normalerweise den Hyaloidearest der Hinterkapsel umziehende weiße Bogenlinie. Arch. f. O. G. **100.** 349. (1919.)

70) Seefelder, R. Beiträge zur Histogenese und Histologie der Netzhaut, des Pigmentepithels und des Sehnerven. Arch. f. O. G. **73.** 527. (1910.)

71) Stähli, Jean. Über Flocculusbildung der menschlichen Iris. Ges. der Schweiz. Augenärzte 1920. Kl. M. f. A. **65.** 107. (1920.)

72) Vogt, Alfred. Die Diagnose der Cataracta complicata bei Verwendung der Gullstrandschen Spaltlampe. Kl. M. f. A. **62.** 593. (1919.)

73) Hesse, R. Zur Entstehung der Kontusionstrübung der Linsenvorderfläche (Vossius). Z. f. A. **39.** 195. (1918.)

74) Fuchs, E. Über traumatische Linsentrübung. Wien. Klin. Wochenschr. Nr. 3 u. 4, 1888, comp. Landsberg, Gunn u. a.

75) Meier, Ernst Albert. Experimentelle Untersuchungen über den Mazerationszerfall der menschlichen und der tierischen Linse. Z. f. A. **39.** (1918) und Inaug.-Diss., Basel 1918.

76) Vogt, Alfred. Zu den von Koeppe aufgeworfenen Prioritätsfragen, zugleich ein kritischer Beitrag zur Methodik der Spaltlampenmikroskopie. Kl. M. f. A. Aug.-Sept. 1920.

77) Vogt, Alfred. Neue Beobachtungen über die Altersveränderungen der menschlichen Linse, insbesondere über die Entwicklung der Alterscataract. Ges. d. Schweiz. Augenärzte, Korrespbl. f. Schweizer Ärzte Nr. 16, 1917 und Kl. M. f. A. **58. 579.** (1917.)

78) Weißenbach, Karl. Untersuchungen über Häufigkeit und Lokalisation von Linsentrübungen bei 411 männlichen Personen im Alter von 16 bis 26 Jahren. Kl. M. f. A. **59.** Nov.-Dez. 1917. Inaug.-Diss., Basel 1917.

79) Krenger, Otto. Untersuchungen über Häufigkeit und Lokalisation von Linsentrübungen bei 401 Personen von 7 bis 21 Jahren. Ein Beitrag zur Kenntnis des Cataractbeginns. Kl. M. f. A. **60.** Febr. 1918. Inaug.-Diss., Basel 1918.

80) Horlacher, Jakob. Das Verhalten der menschlichen Linse in bezug auf die Form von Alterstrübungen bei 166 Personen im Alter von 41˙ bis 83 Jahren. Z. f. A. **40.** 1918, und Inaug.-Diss., Basel 1918.

81) Vogt, Alfred. Faltenartige Bildungen in der senilen Linse, wahrscheinlich als Ausdruck lamellärer Zerklüftung. Kl. M. f. A. **60.** 34. (1918.)

82) Vogt, Alfred. Vergleichende Untersuchungen über moderne fokale Beleuchtungsmethoden. Schweiz. med. Wochenschr. Nr. 29, 1920, S. 613.

83) Van der Scheer J. M. Cataracta lentis bei mongoloider Idiotie. Kl. M. f. A. **62.** 155.

84) F. Pearce, R. Rankine and A. Ormond. Notes on twenty-eight cases of mongolian Imbeciles. B. M. J. 1910. II. Juli, p. 187.

85) Fleischer, B. Über die Sichtbarkeit der Hornhautnerven. Vers. O. G. Heidelberg 1913, S. 232.

86) Leeper, B. Mongols Review of Neurology and Psychiatry Vo. X. 1912, p. 11.

87) Vogt, Alfred. Der Altersstar nach Handmann. Kl. M. f. A. **63.** 397. (1919.)

88) Hippel, E. v. Über experiment. Erzeugung von angeborenem Star bei Kaninchen etc. Arch. f. O. G. **65.** (1907.)

89) Vogt, Alfred. Experimentelle Untersuchungen über die Durchlässigkeit der durchsichtigen Medien des Auges für das Ultrarot künstlicher Lichtquellen. Arch. f. O. G. **81.** 155. (1912.)

90) Vogt, Alfred. Einige Messungen der Diathermansie des menschlichen Augapfels und seiner Medien, sowie des menschlichen Oberlides, nebst Bemerkungen zur biologischen Wirkung des Ultrarot. Arch. f. O. G. **83.** 99. (1912.)

91) Reichen, Jürg. Experimentelle Untersuchungen über Wirkung der ultraroten Strahlen auf das Auge. Z. f. A. **31.** (1914) and Inaug.-Diss., Basel 1914.

92) Vogt, Alfred. Experimentelle Erzeugung von Cataract durch isoliertes kurzwelliges Ultrarot, dem Rot beigemischt ist. Ges. d. Schweiz. Augenärzte. Kl. M. f. A. **63.** 230. (1919.) Further: Schädigungen des Auges durch kurzwellige ultrarote Strahlen, denen äußeres Rot beigemischt ist. Vers. d. Schweiz. Naturforschenden Ges., Lugano 1919.

93) Vogt, Alfred. Experimentelle Depigmentierung der lebenden Iris (Pigmentstreuung in die Vorderkammer) durch isoliertes kurzwelliges Ultrarot, dem Rot beigemischt ist. Ges. d. Schweiz. Augenärzte. Kl. M. f. A. **63.** 232. (1919.)

94) Augstein, Carl. Pigmentstudien am lebenden Auge. Kl. M. f. A. **50.** 1. (1912.) Ferner: Vers. deutscher Naturforscher und Ärzte, Breslau 1904.

94a) Vossius, A. Über Pigmentverstreuung auf der Iris, Hornhaut und Linse etc. Zentralbl. f. pr. A. 1910, S. 257.

95) Axenfeld, Th. Über besondere Formen von Irisatrophie, besonders über die hyaline Degeneration des Pupillarsaums etc. 37. Vers. O. G., Heidelberg 1911, S. 255. Further: ibidem 39. Vers. 1913.

96) Höhmann. Über den Pigmentsaum des Pupillarrandes, seine individuellen Verschiedenheiten und vom Alter abhängigen Veränderungen. Arch. f. A. **72.** 60.

97) Koeppe, Leonhard. Über die Bedeutung des Pigments für die Entstehung des primären Glaukoms und über die Glaukomfrühdiagnose mit der Gullstrandschen Nernstspaltlampe. Arch. f. O. G. **92.** 341. (1916.) Further: Weitere Erfahrungen über die an der Nernstlampe zu beobachtende glaukomatöse Pigmentverstäubung im Irisstroma etc. Arch. f. O. G. **97.** 34. (1918.)

98) Soewarno, M. G. Drei Formen von Irisdepigmentierung. Kl. M. f. A. **63.** 285. (1919.)

99) Elschnig und Lauber. Über die sog. Klumpenzellen der Iris. Arch. f. O. G. **65.** (1907.)

100) Vogt, Alfred. Die Tiefenlokalisation in der Spaltlampenmikroskopie. Z. f. A. **43.** 393. (Festschrift für Kuhnt.) 1920.

101) Seefelder, R. Kammerbucht, Hdb. Graefe-Saemisch, II. Aufl.

102) Erggelet, H. Klinische Befunde bei fokaler Beleuchtung mit der Gullstrandschen Nernstspaltlampe. Kl. M. f. A. **53.** 449. (1914.) Further: Bemerkungen über die Wärmeströmungen in der vordern Augenkammer, ibidem **55.** 229. (1915.)

103) Vogt, Alfred. Die Diagnose partieller und totaler Vorderkammeraufhebung mittelst Spaltlampenmikroskop. Z. f. A. 1920.

104) Koeppe, Leonhard. Zitat 61, further Arch. f. O. G. **97.** 198. (1918.)

105) Koby, F. Ed. Recherches cliniques sur le corps vitré au moyen du microscope binoculaire avec éclairage de Gullstrand. Rev. Gen. d'ophth. April 1920, p. 160.

106) Fuchs, Ernst. Zur pathologischen Anatomie der Glaskörperblutungen. Arch. f. O. G. **99.** 202. (1919.)

107) Fuchs, Ernst. Zur Anatomie der Pinguecula. Arch. f. C. G. **37.** 143.

108) Fuchs, Ernst. Erkrankung der Hornhaut durch Schädigung von hinten. Arch. f. O. G. **92.** 145. (1916.) Further: Über Faltung und Knickung der Hornhaut, ibidem **96.** 315. (1918.)

109) Liebreich. On defects of vision in painters. Macmillans magazine April 1872. Nature Vol. V. p. 404, 506. Brit. Med. Journ. I. p. 271, 296, 318.

110) Heß, C. v. Messende Untersuchungen über die Gelbfärbung der menschlichen Linse etc. Arch. f. A. **63.** u. **64.** (1909.)

111) Vogt, Alfred. Herstellung eines gelbblauen Lichtfiltrates etc. Arch. f. O. G. **84.** 293. (1913.)

112) Saemisch, Th. Im Hdb. Graefe-Saemisch, II. Aufl. Erkrankungen der Lider und der Conjunctiva.

113) Vogt, Alfred. Reflexlinien und Schattenlinien bei Descemetifaltung. Ges. d. Schweiz. Augenärzte 1920. Kl. M. f. A. **65.** 102. (1920.)

114) Vogt, Alfred. Eversion des retinalen Irisblattes. Ges. d. Schweiz. Augenärzte 1920. Kl. M. f. A. **65.** 102. (1920.)

115) Stargardt, K. Über Pseudotuberkulose und gutartige Tuberkulose des Auges. Habilitationsschrift Kiel, Leipzig. Engelmann 1903.

116) Schleich, G. Sichtbare Blutströmung in den oberflächlichen Gefäßen der Augapfelbindehaut. Kl. M. f. A. März 1902, S. 177.

117) Coats. Varicose veins of the conjunctiva. Transact. of the ophth. Soc. 1908, p. 73.

118) Augstein, Karl. Gefäßstudien a. d. Hornhaut und Iris. Z. f. A. VIII. 317. (1902.)

119) Coccius, Adolf. Über die Ernährungsweise der Hornhaut und die serumführenden Gefäße im menschlichen Körper. Leipzig 1852, S. 165, 166.

120) Donders, F. C. Über die am Augapfel äußerlich sichtbaren Blutgefäße. Vers. O. G., Heidelberg, II., comp. further 3. Jahresbericht d. Utrechter Augenklinik.

121) Friedenwald. Der sichtbare Blutstrom in neugebildeten Hornhautgefäßen. Zentralbl. f. pr. A. Jahrg. 1888, S. 32.

122) Elschnig, Anton. Über den Keratokonus. Kl. M. f. A. **32.** 25. (1894.)

123) Krückmann, E. Die Erkrankungen des Uvealtractus. Graefe-Saemisch, Handb. 1907.

124) Uhthoff, W. Über einen Fall von Keratokonus mit Sektionsbefund. Kl. M. f. A. Beilageheft 1909, S. 41.

125) Hedinger und Vogt. Klinische und anatomische Beobachtungen über Faltung der Hornhaut, der Linsenkapsel und der Retinaoberfläche. Arch. f. O. G. 102. 354. (1920.)

126) Hirschberg, Julius, named the flocculi of the human eye Ectropium uveae congenitum. Comp. Ancke, C. f. A. 1885. S. 311—313 and ibidem J. Hirschberg, 1903: „Über angeborene Ausstülpung des Pigmentblattes der Regenbogenhaut". Comp. Colsmann, Kl. M. f. A., 1869, p. 53, Holmes, E. O., Chicago, Med. Journ., June 1878, u. a.

127) Stähli, Jean. Über Flocculusbildung der menschlichen Iris. Kl. M. f. A. 65. 349. (1920.)

128) Gullstrand, Allvar. Die Dioptrik des Auges. Hdb. d. physiologischen Methodik von Tigerstedt. 1914.

129) Graefe, Albrecht von. Über essentielle Phthisis bulbi. Arch. f. O. G. 12. 261. (1866.)

130) Gilbert, W. Über chron. Uveitis und Tuberculide der Regenbogenhaut. Arch. f. A. 82. 179. (1917.)

131) Stock, W. Tuberkulose als Ätiologie der chronischen Entzündungen des Auges etc. Arch. f. O. G. 66. (1907.)

132) Igersheimer, Josef. Die ätiologische Bedeutung der Syphilis und Tuberkulose. Arch. f. O. G. 76. (1910.)

133) Hippel, E. v. Über tuberkulöse, sympathisierende und proliferierende Uveitis unbekannter Ätiologie. Arch. f. O. G. 92. 421. (1917.)

134) Koeppe, Leonhard. Arch. f. O. G. 92. 115. (1916.) (Über Iritis tuberculosa nebst Bemerkungen über therapeutische Erfolge durch Bestrahlung mit der Lampe.)

135) Streiff, J. Zur method. Untersuchung der Blutzirkulation in der Nähe des Hornhautrandes. Kl. M. f. A. 53. 395. (1914.)

136) Bajardi, P. Sull esame microscopico della circolazione dei vasi della congiuntiva umana. Congr. oft. di Palermo 1892.

137) Bajardi, P. Ancora sull esame microscopico dei vasi della congiuntiva nel vivo. X. Congresso d'oftalmologia. Lucerna 1904.

138) Koeppe, Leonhard. Arch. f. O. G. 97. 9ff. (1918.)

139) Koeppe, Leonhard. Arch. f. O. G. 93. (1917.)

140) Elschnig, Anton. Klinisch-anatomischer Beitrag zur Kenntnis des Nachstars. Kl. M. f. A. 49. 444. (1911.)

141) Gjessing. G. A. Harald. Kliniske Linsestudier, Drammen 1920.

142) Fleischer, B. Über Myotonia atrophicans und Cataract Vers. O. G. Heidelberg 1916, comp. J. G. Greenfield, Rev. of Neurology and Psych., IX. 169, J. Hoffmann, Arch. f. O. G. 81. 12. (1912), A. Hauptmann Kl. M. f. A. 60. 576. (1918.)

143) Tscherning. Optique physiologique (1898) p. 41.

144) Fridenberg, P. Über die Figur des Linsensterns beim Menschen und einigen Vertebraten. Arch. f. A. 31. 293. (1895) and Arch. of ophth. April 1895 (The lens star figure of man and the vertebrates). Comp. Carl Großmann, Intern. Ophthalmologen-Kongreß Luzern 1904.

145) Goldberg, Hugo, Pigmentkörperchen an der Hornhauthinterfläche. Arch. f. A. 58. 324. (1907.)

146) Kraupa, Ernst, Studien über die Melanosis des Augapfels. Arch. f. A. 82. 67. (1917.)

147) Verderame, Ph., Visibilité des nerfs cornéens à l'état pathologique. Rev. Gen. d'ophth. 34. 505. (1920.)

TABLE OF CONTENTS

(The numbers given refer to the pages)

Arranged by DR. U. LÜSSI.

Printed by F. A. Brockhaus, Leipzig.